Agrofuels

Transnational Institute

Founded in 1974, the Transnational Institute (TNI) is an international network of activist-scholars committed to critical analyses of the global problems of today and tomorrow. We seek to provide intellectual support to those movements concerned to steer the world in a democratic, equitable and environmentally sustainable direction.

In the spirit of public scholarship, and aligned to no political party, TNI seeks to create and promote international co-operation in analysing and finding possible solutions to such global problems as militarism and conflict, poverty and marginalisation, social injustice and environmental degradation.

Email: tni@tni.org
Website: www.tni.org
Telephone + 31 20 662 66 08
Fax + 31 20 675 71 76

De Wittenstraat 25
1052 AK Amsterdam
The Netherlands

AGROFUELS

Big Profits, Ruined Lives and Ecological Destruction

François Houtart

Translated by Victoria Bawtree

With the collaboration of Bosco Bashangwa Mpozi,
Bienvenue Lutumba Bukassa and Geoffrey Geuens

Foreword by Walden Bello

PlutoPress
www.plutobooks.com

First published in French as L'agroénergie: Solution pour le climat ou sortie de crise pour le capital by Couleur Livres (Charleroi, 2009)

First English language edition published 2010 by Pluto Press
345 Archway Road, London N6 5AA and
175 Fifth Avenue, New York, NY 10010

www.plutobooks.com

Distributed in the United States of America exclusively by
Palgrave Macmillan, a division of St. Martin's Press LLC,
175 Fifth Avenue, New York, NY 10010

British Library Cataloguing in Publication Data
A catalogue record for this book is available from the British Library

ISBN   978 0 7453 3013 6   Hardback
ISBN   978 0 7453 3012 9   Paperback

Library of Congress Cataloging in Publication Data applied for

This book is printed on paper suitable for recycling and made from fully managed and sustained forest sources. Logging, pulping and manufacturing processes are expected to conform to the environmental standards of the country of origin.

10   9   8   7   6   5   4   3   2   1

Designed and produced for Pluto Press by
Chase Publishing Services Ltd, 33 Livonia Road, Sidmouth, EX10 9JB, England
Typeset from disk by Stanford DTP Services, Northampton, England
Printed and bound in the European Union by
CPI Antony Rowe, Chippenham and Eastbourne

Contents

Foreword

Walden Bello

Francois Houtart has always been on the cutting edge of critical research and analysis. This book reveals that his indefatigable mind has lost none of its capacity for sharpness, comprehensiveness, and originality. This time, he brings his formidable strengths to deconstructing the agrofuels bonanza. The result is a work that marshals a wide array of facts into an expose and places that expose in a paradigm with tremendous explanatory power.

Agrofuels, Houtart argues impeccably, are not what they're cracked up to be. Their advantages over fossil fuels are oversold. They have negative social, economic, and ecological side effects that outweigh their positive impacts. All this Houtart documents with admirable detail.

Yet why do agrofuels continue to be promoted as one of the solutions to climate change? Here is where Houtart's analytical strengths come into play. He explains this conundrum by locating the agrofuels bonanza in the twin crises of energy and climate and locating the latter in turn in the dynamics of capitalism – in capitalism's insatiable drive to convert living nature into dead commodities, a process that is governed by the logic of the market and propelled by the search for profit. Fossil fuels, he explains, are central to the rise to hegemony of corporate capitalism, so that one cannot see global warming as simply a byproduct of a certain technology but as an inevitable outcome of a certain social organization of production.

It is not coincidental that the energy and climate crises are coming to a head at a time that the reigning ideology has been neoliberalism, with its worship of the market and its opposition to the regulatory role of the state. Appealing to the market as the spur of efficient production and distribution, corporate capital has created a singularly vulnerable socioecological system resting on the profligate extraction of fossil fuels, overconsumption, and massive waste.

As climate change has become a reality, the development of alternatives to fossil fuels has taken place within the coordinates of the dominant paradigm, following the logic of the market and profitability. Herein lies the answer to the conundrum of agrofuel's limited potential and the hype with which it is being promoted. Under the neoliberal capitalist paradigm, agrofuels are made profitable for corporations via subsidies despite the dislocations they create; carbon markets are preferred to carbon taxes; and nuclear power is made respectable – and profitable – once more despite its unresolved dangers. Despite the fact that market failures have dragged the world to the deepest downturn since the Great Depression, we are being made to place our trust in the magic of the market and the spur of profit to find ways to meet the challenge of global warming. This is, Houtart says, a fool's errand.

Houtart does not confine himself to decrying the tragedies inflicted by capitalist development. He looks optimistically at the prospects of renewable energy sources such as solar and wind. The obstacles to the development of alternatives, he makes clear, are not technical blocks but social ones. Unless alternative energy sources and systems are developed within what he calls a post-capitalist system, there will be no viable long-term solutions to the twin crises of the climate and energy.

Few can deliver this message with that rare combination of impressive research, solid analysis, and controlled passion. Francois Houtart is one of them.

Walden Bello is a member of the House of Representatives of the Republic of the Philippines. Formerly a professor of sociology at the University of the Philippines and executive director of Focus on the Global South, he is the author or co-author of 15 books, the latest of which is *Food Wars* (London and New York: Verso, 2009). He is a Fellow of the Transnational Institute.

Preface

The question of biofuels has become an ideological problem, a concept that requires thorough rethinking. To say it in more technical terms, it is a unique signifier that has changed significance. There was a time when to be in favour of biofuels was an ecological position and rather leftwing, because 'bioenergy' was believed to correct the defects of fossil energy. In contrast, the right saw it only as an unrealistic environmental dream or a veiled criticism of the growth created by the capitalist economic system.

Now things have changed. It is more the right that defends biofuels and the left that attacks them. This is partly because the energy and climate crises have become unavoidable realities and can no longer be ignored. It is also because the search for new sources of energy, caused by the price of oil and gas, has become a very profitable activity for investors with capital. Bio- or agrofuels were looked on favourably by public opinion, increasingly aware of the environment problem. However, such economic reasoning does not take into account the 'externalities', i.e. the ecological and social conditions in which the new fuels are produced and their effects on nature and populations.

It is this last aspect that is being emphasized by the social movements today. They know that the capitalist system usually makes its economic calculations for the short term and that it ignores the real cost of what it believes to be external to its own logic, and that constitutes collateral effects. The whole question of biofuels is thus being seriously reconsidered.

As a result, there is an ideological war going on, which is fought with words. Both sides make their arguments loud and clear. Some of them highlight the advantages of biofuels, the efforts being made to save energy and the transformation of large oil, industrial and commercial groups into genuine benefactors of humanity. Indeed, all dressed in green, they invoke the immense possibilities of science and technology which, according to them, will resolve in the foreseeable future the issues that are still pending today – on condition that private enterprise is left free to become involved, without hindrance, in this new profitable market. The case of Senator McCain in the United States is exemplary. In 2000 he violently criticized ethanol, calling it 'an agro-industrial boondoggle', and in 2006 he called it a real source of energy for the future.[1]

However, the social movements, the leftwing parties and a certain number of progressive non-governmental organizations (NGOs) refuse to use the term 'biofuels', preferring the more descriptive expression 'agrofuels', less linked to the optimistic connotation of 'bio' (life). Some of them go as far as proposing the term 'necro-[death] fuels'. Their association with the food crisis and the image of full granaries as opposed to empty plates has gained ground.

These semantics have invaded the precincts of the United Nations (UN), the Food and Agriculture Organization (FAO) and the World Trade Organization (WTO). On the one hand, the need for publicity ends by deforming the meaning of the words and presenting measures that are simply corrective to the preceding destructive practices as advances in the progress of humanity. On the other hand, the arguments of those who have observed the ecological and social disasters, not only of fossil energies but also of how certain renewable energies are produced following the logic of economic interests, are sometimes too simplistic or they ignore certain technical aspects of the problems. Sometimes shortcuts are made in the links between effects and causes that weaken their positions.

For us there is no question of ignoring the ecological and social 'externalities' in considering the double crises of energy and climate and the question of new energies, particularly from agrofuels. This will inevitably lead us to a critique of the dominant economic discourse, because it overlooks an essential part of the reality. Ours is not an apocalyptic discourse, far from all hope of solutions, including those in the scientific and technical fields, but neither does it ignore the profound gravity of the situation or the duplicity of reassuring discourses. Finally, it is a question of not being satisfied with slogans that purportedly serve the cause of the victims of a system, if such slogans have no scientific or logical base.

However, this work is not neutral. It is part of the pursuit of justice and the construction of an economic and political logic that is respectful of ecological equilibrium and human welfare. The aim is to take an ethical approach, in defence of life, and not to conceal indignation about what is, for many, a death sentence. This approach will be based on history and take into account all the different situations, including what makes it possible to analyse them as a whole. In this way it is easy to legitimize one's own position, as is done by the economic reasoning of capitalism, which ignores externalities. The social reality will be analysed as the result of actors interacting, that is to say, not as a linear process but a dialectical one, in which bargaining powers come into play to transform and build social structures or to halt their transformation.

The problem of agrofuels, as we shall see, lies at the very heart of social relationships, because energy is the pivot of the capitalist market economy and even of what is called 'Western civilization'. It is the reason why the economic and political powers tend to adopt solutions that enable the development model to continue, without challenging its parameters. The whole question is, therefore, to know whether such logic is achievable and at what price, or whether, on the contrary, another logic should underpin humanity's future.

This book aims at describing the double crises of energy and climate, and then posing the question of new energies, particularly agrofuels. It closes with a reflection on the real functions of this new production and the need for radical solutions if humanity is to extricate itself from the impasse in which it now finds itself.

A number of people have assisted in this work with their specific expertise and I would like particularly to acknowledge the contributions of Bosco Bashangwa Mpozi, a bio-engineer who teaches at the Institut des Techniques de Développement (ISTD, Mulungo, Democratic Republic of Congo) and of Bienvenue Lutumba Bukassa, agro-chemist, assistant and researcher at the Institut Facultaire des Sciences Agronomiques de Yangambi at Kisangani (Democratic Republic of Congo). Both of them were associate researchers at the Tricontinental Centre during the preparation of this book. Eric Feller, agronomist and researcher at the University of Liège, contributed with his technical advice, and Geoffrey Geuens, professor in communications at the University of Liège, undertook a study on the multinationals involved in this field, which has served as a base for part of the economic section. Many thanks, too, to Leonor García for her efficiency in presenting and laying out the manuscript, as well as to the French Catholic Committee against Hunger and for Development (CCFD) and Christian Aid of the United Kingdom for the support they have given to this work.

1
Energy and Development

These days agroenergy, or green energy, is being touted as a solution for the future. The facts about global warming, particularly the dramatic increase of carbon dioxide (CO_2) in the atmosphere, have made people realize that action must be taken. True, these two phenomena are not only linked to energy problems. The production of greenhouse gases by certain forms of agriculture is also considerable, above all because of the increase in livestock production. However, the question of energy lies at the heart of the issue, particularly in the industrially-developed countries, both for industrial production as well as for heating and transport.

This is what led the European Commission (EC) to put forward measures to the member states of the Union. Since March 2007 the objective has been to reduce greenhouse emissions by 20 per cent by 2020 compared to 1990 levels, and by 30 per cent if global agreement is reached that renewable energies should constitute 20 per cent of energy use, with 10 per cent of transport operating on agrofuels (the 10 per cent target for agrofuels was later reduced to 8 per cent in response to criticism). In January 2008, the EC proposed a climate energy packet to each member state, according to its wealth, which included, *inter alia*, the burying of CO_2 in old mines and the constitution of a new European emissions trading system. Apart from industry, sectors like habitat, agriculture and transport should reduce their carbon emissions by 10 per cent by 2020, in comparison with 2005. All this was estimated to cost some 60 billion euros a year. Such an objective seemed ambitious at the time but, as we learnt afterwards, it risked being far below the real needs for effective action to save the planet. In December 2008 the 20-20-20 plan was adopted: 20 per cent fewer CO_2 emissions, 20 per cent less consumption and 20 per cent more renewable energy, by 2020.

Fossil energy is being challenged because it is non-renewable and polluting. Research for alternatives is under way, but it is far from being disinterested. In fact, various interests have become involved in the desire to produce according to a so-called 'sustainable' model, without penalizing future generations. This is why the nuclear industry does not hesitate to present itself as a solution, even though it is based on uranium, a non-renewable raw material. Furthermore, the problem of nuclear waste has in no way been resolved. As for the question of oil and its replacement, this is also linked to geopolitics: consider, for example, the dependence of the United States on oil from the Middle East and Venezuela. In the former, this led to the war on Iraq and Afghanistan and in the latter to an ethanol partnership proposed by President George W. Bush to President Lula of Brazil, which shows how the US is trying various solutions to get round the problem. At the outset of the twenty-first century, their countries are the two largest producers of agroenergy.

Whole populations have been suffering from ecological damage for a long time now. As long as it was the lower social classes or colonized peoples, the economic and political powers in the industrialized countries were hardly concerned about the problem. Since the beginning of the industrial revolution the places where production was concentrated, which were also the places where the working classes lived, became highly polluted. Havoc was wreaked on the countryside, forests and also the habitats of colonized peoples by the exploitation of natural resources. Few denounced this, as it was the price of progress. The situation had to get a lot worse, to the point of affecting economic interests and the quality of life for all sectors of society, including the dominant classes, for ecological destruction to become an issue. This is why the question of agroenergy has now emerged as one of the political priorities.

To avoid a one-sided view, it is essential to develop a historical and universal understanding of the question. Interest in agroenergy did not fall from the skies. It should be seen as part of a long process of the exploitation of nature without great concern about its ability to reproduce itself, which is linked to contempt for the working classes, and for the peoples living on the periphery. Energy has been a human need at all times: it could be said that the history of humankind has run parallel to that of the utilization of energy. Being both product and cause of technologies, the development of energy use has enabled personal mobility – and therefore transport – to be extended. It is one of the fundamental aspects of what we today call globalization, which is characterized by the liberalization of trade, developed on the principles of capitalism. Capital, considered to be the engine of development, has been able to construct the foundations of its own reproduction as a world system, thanks to the new information and communication technologies. Energy has played a key role in the process, as it is at the heart of the two main activities of the economy: production and transport. Both have considerably increased during the neoliberal phase of capitalism, i.e. with the generalized liberalization of trade. Demand for energy has rocketed – with all the consequences we know.

The mode of living which has resulted is particularly energy-hungry. At the beginning of the modern age, energy was abundant and its cost was very low for a long time, so its utilization was virtually limitless in the industrialized world. Until the day came when the destructive effects of such practices endangered the development model itself, not only because certain resources were becoming exhausted, but also because of its ecological and social effects. The 'Cry of the Earth' joined with the 'Cries of the Oppressed' and it was no longer possible to ignore them. The convergence between the two was to shape the resistance to the neoliberal model.

The West's destructive exploitation of the environment has been encouraged by some of the attitudes and ideals that have been nurtured within Western cultures in the modern age. It is believed by many that Western civilization is on a constant forward march, progressively advancing in prosperity and technology. The natural world is to be tamed, exploited and transformed in the service of humanity's advancement. Any idea that this march of progress might be permanently halted, or that there might be some limit to Nature's bounty on which our existence ultimately depends, are put to one side, so that the vision of the future is one of a constant upward trajectory of economic growth. The achievements and language of Science have helped nurture these conceptions, and in the theory of evolution there appeared to be some scientific backing for this notion of Progress.

This vision of the world developed within social relationships that were highly unequal, both between social classes and between the peoples of the world. It gradually became the ideology of the dominant groups, that is to say, both the explanation of their 'avant-gardism' and the justification for their place in society. The role of capital, as the promoter of progress and bearer of hope for the future, was both real and illusory. On the one hand, thanks to the logic of accumulation and profit, following the laws of a market subjected to this rule, the production of commodities and services grew at an unprecedented rate. In its neoliberal phase, the acceleration was ever more spectacular. During the second half of the twentieth century, global wealth multiplied sevenfold. On the other hand, the process was also illusory, because it hid several realities: the way in which the production was achieved, the subsequent distribution of wealth and the destruction of the environment.

In fact, the method of production prefigured future ecological catastrophes and social disasters. As for the distribution of wealth, it has ended up in a process of concentration and exclusion – as a logical consequence of capitalism. The latter gives priority to exchange value over use value, thus subjecting economic activity and that of numerous sectors of the common good to the law of the market, considered as natural and prevalent.

Ignorance about what are called 'externalities', i.e. the factors that are not taken into account in economic calculations, was to lead to very serious contradictions. This situation is caused by not taking into account the ecological and social costs of production and transport in the initial stage and the unequal distribution of the products at a subsequent stage. But this is not the simple result of some natural law or the price to pay for progress. It corresponds to the very specific interests of certain social classes, linked to capital accumulation, having every advantage in maintaining a high rate of accumulation and little concern for what one could call the common good. Thus it is not surprising that mainstream economics has taken the form that it has, notably excluding social and environmental externalities.

The social and ecological crisis has become so great that nobody can ignore it any longer. It is even affecting the rate of profit and hence the interests of capital, whose reproduction is being threatened, and there is a risk of world economic stagnation.

Solutions must therefore be found. According to the logic of capital, which had found a new dynamism in the development of liberalized trade, such solutions must be found within the continuity of the system. It is thus a question of implementing alternatives, changing certain behaviour, but in no way challenging the logic of capitalist accumulation. This is always put forward as the necessary solution, as long as it accepts certain changes and regulations.

A typical example of this perspective is the film *An Inconvenient Truth* by Al Gore, Nobel Peace Prize winner. The film rightly put its finger on the world ecological problem, shook up public opinion and was favourably received in liberal economic and political circles. When the former US Vice-President visited Belgium, it was not the socialist party or the heirs to Christian democracy who received him but the liberal francophone party. The reason is simple:

Al Gore's film does not question the system. He sees the solution as being essentially a question of individual behaviour: reducing the consumption of electricity, moderate use of automobiles, installing double glazing, etc. His speech is moralizing and he even makes use of religious arguments. He addresses individuals and not social mechanisms for the transformation of the economic model.

Energy is obviously central in this whole issue. It is at the heart of the capitalist development model itself, for without recourse to these energy resources the system would not function any more. Hence, if the existing forms of producing energy are in contradiction to the reproduction of the economic model and of society, new ones must be sought. This is where the development of agroenergy comes in, as a substitute for fossil energy. But the question is whether this is a solution or a palliative, and to find this out it is necessary to examine it more closely.

ENERGY IN THE DEVELOPMENT MODEL

Without energy there is no development, so that the two realities are in fact but one. The history of the former cannot be written without involving the history of the latter. It is not only a material question but also one of the cultural meshing which inevitably takes on political dimensions. The utilization of energy is therefore an integral part of what we might call human dynamism. The different phases of the history of humanity are clearly marked by the use of various sources of energy.

According to the French sociologist and philosopher, Edgar Morin, one paradigm has always directed the course of the physical, biological and anthropological world: the reorganization of life. Life is reproduced in a series of sequences, going from order to chaos and from disorder to reconstruction. A vital elan certainly marks a chaotic evolution, but when it is a question of human beings, there is an enormous capacity for inventiveness. Energy has contributed greatly to this inventive capacity.

Transformations have also taken place in the field of relationships with nature, as well as of social relations. In the former case, human beings gave proof of a constant capacity for adaption, passing from simple hunting and gathering to the organizing of agriculture and then moving gradually to trading and industrial societies. As for social relations, inequality has usually prevailed. Male domination is at the base of the distribution of roles between men and women in domestic life, as well as in the economic, political, cultural and religious fields. The possibility of access to non-manual work, often linked with racial belonging, defined the castes of the pre-capitalist societies, while work exploitation under industrial capitalism has given rise to the social classes. All through history there have been peoples who became dominant or imperial, imposing their interests on others.

In this process, the control of energy played a very important role. It was at the base of agricultural, artisanal and industrial activity. Greek mythology reminds us of its central place, through Prometheus and the mastering of fire, or Sisyphus and the unending effort to overcome gravity. From the beginning of human history, the utilization of natural energies was introduced as a survival mechanism, whether it be sun, wind or water, but also animal and human energy. Gradually wood, then coal, was transformed into heat, water into steam, oil into petroleum – ending finally with nuclear energy. Nowadays we distinguish between renewable energies and those that are not, i.e. the ones that use raw materials that do not have a cyclical existence. As for polluting sources of energy, their usage is increasingly affecting the atmosphere and even the climate through the noxious emissions of CO_2 or methane, resulting from their combustion, and accelerating the production of noxious micro-particles into the ozone layer.

During the last two centuries of industrial development, the exhaustion of resources was not considered at all. The impression was that the planet had an unlimited capacity to respond to human needs and if, in one region, iron, copper or coal was no longer available, there were always many other places where they could be found in abundance. Besides, the new technologies made it ever more possible to use natural resources better, to increase their profitability and to find new ways of exploiting deposits that had previously been considered unattainable. In fact the notion of endless progress made it possible to envisage scientific advances and their technological application being one day able to resolve problems considered insoluble at the present time.

Energy was part of this same philosophy. Optimism was the rule and nothing seemed able to stop the conquest of humanity whose dynamism developed into a frantic, ever-increasing consumerism. Georges de Cagliari, in his theatrical piece, *Le Feu de la Terre*, expressed this frenzy when he described this period as a 'modern prehistoric era ... without harmony with nature, which no longer exists'. The oil crisis was necessary to alert public opinion about the cost of energy and its non-renewable nature, as well as the Chernobyl disaster, to remind people of the dangers of nuclear energy and show that the advantages of atomic energy were relative. As for acid rain and climate warming, they contributed in an increasingly visible way to remind us that human activity, especially in the field of energy, has consequences that could be catastrophic.

ENERGY'S ROLE IN THE GROWTH OF CAPITALISM

Mercantile societies developed on the basis of trade, which was the result of work and therefore a separate activity from agricultural production. Also they were not able to build themselves up unless their agricultural activities produced enough to feed not only the peasants themselves but others as well. Hence the importance of transport, both of agricultural products towards the towns and the artisanal trading between urban agglomerations. This could not be done without using new sources of energy, particularly that of animals. It should be remembered that the transformations were not only in the energy field. They were also at the origin of a new social organization, the development of politics, the birth of an ethic and finally a new vision of the world. The latter, freeing itself from nature's cycle, led to a notion of progress in time and in space, which also oriented the utilization of energy.

With the development of capitalism, the situation changed profoundly. The trade in merchandise enabled the accumulation of capital which itself became a source of profit and gradually changed into being the engine of the economy and of society. In Europe the phenomenon began at the end of the eleventh century, with the development of trade between East and West, along the rivers, and with the expansion of commercial towns and the development of a bourgeoisie that was based first on trade and later on industry.

Energy played a still more important role in the second phase of capitalist development. In its mercantile phase, it had not produced great energy revolutions. It was concentrated on the extraction of mining and agricultural wealth, which did not need more than animal and human energies. This explains particularly the slave trade that emptied the lands of Africa to replace the populations of pre-Colombian America who were dying out after the Spanish conquest. As for intercontinental transport, it used wind power.

In contrast, industrial capitalism was built on considerable changes in energy. The role of the steam engine, and therefore the great increase in the use of coal, in all fields of production is well known. Economic activity took on a new dimension that gave capital a central role. The unifying role of capitalism has been explained by Marx, when he showed that in the first stage capital was bringing together artisans in factories in one place and in the second stage has organized production through the division of labour where workers were no longer masters of the fruits of their labour. Later on capital monopolized distribution too. As a result there was a veritable explosion in the production of goods and services, an increasing exploitation of nature and growing social differentiation in antagonistic classes. The exploitation of the natural resources of the periphery expanded considerably through colonial enterprises. The intra-European and world wars were the result of ferocious competition to ensure their control.

With the Washington Consensus of the 1970s, a new period dawned, caused by a crisis in the accumulation of capital in which the oil crisis played a role. Neoliberalism, preaching the total liberalization of capital, goods and services (but not workers), aimed at freeing the economy from the obstacles that had been established by the three great models of the post-war period: Keynesianism, socialism and the national development of the Third World countries. In these three cases, a limit was established to the expansion of capitalist accumulation, both by the social pacts redistributing the national wealth between capital, labour and the state, as well as by the development of socialism, the system that was in principle an alternative to capitalism, and finally by the importance given to the state as an engine for industrial development. It was therefore necessary, according to the theses of Friedrich von Hayek and Milton Friedman, to liberate market forces to bring new life to the accumulation of capital necessary for the development of new technologies, especially in the field of information and communication, and thus to respond to the vast needs of the concentration of productive and financial capital created by transnational corporations buying up other corporations.

This project, supported by an international institutional framework, particularly the World Bank and the International Monetary Fund (IMF), led to the reinforcement of the power of the economic decision-making centres of the Triad (the United States, Europe and Japan). It ended in constituting a minority of the world population (some 20 per cent) as hyper-consumers that were particularly energy-greedy. This restrictive model was in fact favourable to the accumulation of capital because it enabled a much faster circulation of capital and a much greater production of value added than on ordinary consumption goods, accessible to the largest number of people; not to mention those who came into the category of the 'useless masses' (for capital), as they did not produce added value and did not dispose of purchasing power that enabled them to accede to the status of consumers.

At what point are we now, at the beginning of the third millennium? The consumption of non-renewable natural resources, particularly energy, by a minority of the world's population would require, according to certain estimates, the reproductive capacity of more than three planets if extended to the world in general. But we only have one planet and we must therefore act quickly. The development model being following by the so-called emerging countries is no different from that developed by the industrial countries. Brazil, for example, which expected to adopt a different economic model, has hardly changed the neoliberal orientation of its economy and allies with the United States in a common front on ethanol, which benefits the large landowners and multinational agrobusiness enterprises, without questioning the consumption model. China and Vietnam have chosen to open up to the capitalist market, which has enabled a spectacular development: 20 per cent of their populations have rapidly reached the level of consumption in the Triad. India also entered into the neoliberal model in the 1990s when it abandoned the project of national development, and it has followed the same logic, but with social differences still greater than they were previously.

All these recent models of development of the periphery show little consideration for the non-renewable nature of fossil fuels. On the contrary, they are competing with Western economies whose advantages in the production of goods and services are evident to them. They are reluctant to take ecological measures of conservation, arguing, not without reason, that their turn has come and that the countries who have been mostly responsible for laying the world to waste have dubious motivations in imposing restrictions upon other countries which they have not been able to respect themselves and which have enabled them to occupy a dominant place in the world economy.

It should be added that demographic trends have considerably aggravated the situation. At the beginning of the twentieth century, there were 1 billion human beings in the world, and at the dawn of the twenty-first century, there were more than 6 billion. There will probably be 9 billion towards 2030. True, the birth rate has tended to go down in the regions as a whole, but the progress that has been made on bringing down mortality rates is largely responsible for this evolution.

The demographic expansion has, in fact, taken place in the framework of capitalist logic, which concentrates wealth and accentuates the gaps between the rich and the poor, especially in the consumption of energy. Neoliberal capitalism has certainly increased the absolute number of wealthy people on the planet. However, at the same time the absolute numbers of those who live in poverty, or even in extreme poverty, also continue to grow. In Latin America, at the beginning of the twenty-first century there were 220 million poor people (according to the World Bank's definition, which is those who earn less than two dollars a day) which means an increase of 20 million poor people in a decade. In July 2008, Jacques Diouf, Director General of the FAO, announced that in 2007 the number of those suffering from hunger in the world had increased by 50 million due to higher food prices. There is therefore a minority of human beings who, by their mode of development and consumption, contribute the most to the negative social and ecological effects of the utilization of energy.

To understand the links between this phenomenon and the logic of the accumulation of capital, we should recall the book by Susan George, *The Lugano Report*. The author imagines the following scenario: several heads of the great multinational corporations, worried about the economic evolution of the world, ask a group of experts to study how to save the capitalist system. The group, after much research and scholarly calculations, come to the conclusion that, in order to do this, half of the world's population should be eliminated, that is to say, the 'useless masses' that do not contribute to the increase of wealth, or to profits that could be made on sales.

Naturally they defended themselves from the accusation of genocide, saying that it could be left to nature, as the endemic diseases, on the one hand, and the capacity of human beings for self-destruction, on the other, would lead to the desired result. Susan George explains in her last chapter that it is a work of fiction, but that such reasoning existed and it demonstrated a certain logic.

Confronted by global demographic trends, the World Bank believes that agricultural methods should be transformed in order to feed the population of the future. To achieve this it has long been preaching that peasant agriculture should be replaced by capitalist productivist cultivation. This is the model of the United States, which has been imposed over the last few decades in certain regions of Latin America, particularly for the cultivation of eucalyptus (for paper and charcoal) and soya (for oil as a possible substitute for fossil energy). But as we shall show later on, *apropos* of oil palms (African palm), this approach is extremely destructive of soils and the quality of water and also requires the destruction of primeval forests, as well as being socially disastrous. The local surplus population is uprooted from its region and sometimes even massacred (as in Colombia) and then concentrated in unsanitary dwellings of the large towns or contributing to swelling international migration pressures.

It should once again be recalled that, in its economic calculations, the capitalist logic that organizes the extraction and utilization of energy sources does not take 'externalities' into account. This is exemplified by a case from Sri Lanka. In 1996 the World Bank told Sri Lanka to abandon rice culture in favour of industrial crops for export. The problem: the cost of rice production was higher than in Vietnam and Thailand. The logic of the market thus demands priority be given to importation. To implement this project the Bank demanded that the Sri Lankan government abolish the government bodies that regulated the rice market, fix a tax on irrigation water so that rice production was no longer profitable and finally give all the small peasants the right to own property. The rice lands were still held in common, as in the old Asiatic mode of production, and they belonged to local collective bodies. Their transformation into commodities enabled the peasants to sell their land at a low price to local and international companies capable of undertaking the new type of production, which was mainly destined for export, such as crops producing agrodiesel and ethanol and, at the beginning, sugar cane. After some hesitation, the Sri Lankan government produced a document entitled *Regaining Sri Lanka*, stating that the idea was not bad and would enable the country to provide cheap labour to attract foreign capital.

After this policy had been carried out by the country over some 40 years, in the form of free trade zones, the workers had been able to push up the level of wages, organize a form of social security and establish a pension scheme. In brief, labour had become more expensive in Sri Lanka and already some foreign investors had left the country for Vietnam or China, where wages were lower. The conclusion of the government was that the price of labour had to be lowered and therefore the real wage reduced, and certain aspects of social security had to be dismantled as well as the level of pensions. This was the result of the logic on which the whole policy was based. Such economic reasoning takes no account of factors that are not involved in the market calculation, such as food sovereignty[1] (Sri Lanka being an island), the well-being of a million small peasant rice producers, the living standards of industrial workers, the quality of food (the type of rice varies in the different countries), the energy costs of transport – not to mention the history and culture of the rural world. There is no question of including externalities in economic calculations: the capitalist market logic is implacable and it is the only one that is taken into consideration in the neoliberal organization of the world economy.

It is the same in the field of energy, where the national and social conditions of exploitation started to be taken into account in economic calculations the day when scarcity became a reality, when the oil states pushed up the bidding and when workers in the different fields of energy organized to obtain better working conditions and wages. Ignorance about the national and social factors, as was the case for the exploitation of coal, oil and gas, risks being repeated in agroenergy, if the alarm bell is not sounded in time.

THE SOCIAL AND ECOLOGICAL EFFECTS OF THE CAPITALIST DEVELOPMENT MODEL

It took some time for it to be generally recognized that the model of production and consumption pushed to its extreme by neoliberal logic was no longer tolerable. In the 1950s the Club of Rome advocated zero growth. This seemed out of keeping with the economic euphoria of that time, in contradiction with all the paradigms, above all those that held progress to be linear and that it was possible for science and technologies to resolve all economic and social contradictions as and when they arose. Also, when a number of social sectors had, for the first time, reached a certain level of consumption, the message was politically unacceptable. Zero growth seemed to be a return to the past, a denial of the right to development – a veritable regression.

The authors of the document no doubt did not insist enough that zero growth did not mean a drop in the quality of life. In fact, their position was that the same level could be ensured by a less brutal use of natural resources, particularly energy.

The development of ecological movements and their emergence into the political arena were also important factors for a growing awareness of the need for basic changes in lifestyles in the industrialized countries. They drew attention to phenomena that were immediately visible, such as the disappearance of certain animal species, the damage being done to forests, the destruction of the soil, air and water pollution, and the irrational use of energy, particularly in the means of transport. Their many campaigns alerted public opinion to the point of forcing all political parties to put the problem on their agenda.

The United Nations Conference on Sustainable Development, held in Stockholm in 1987, was an important date in this process. It adopted a new concept, which became a standard UN phrase and was finally adopted internationally. This definition introduced the notion of the future of the generations to come. A sustainable development is one that does not put the future in danger and therefore makes it possible, while using natural resources and especially energy, to conserve the universal heritage for it to be used later. The great weakness of the position adopted was not to pose the phenomenon in relation to the economic development model. The report of the Brundtland Commission, which bore the name of its president, the former prime minister of Norway, indicates that precautionary measures were indispensable, but within a model of economic development that was not questioned and whose logic finally advocated the contrary.

More global attitudes, attacking the roots of the problem, began to emerge in critical literature, but also at the beginning of certain social movements, such as the coordination of peasant movements into the international movement, La Via Campesina. The world of the peasants, the primary victim of the capitalist world in the field of agricultural production, was logically the most appropriate to warn the public. Later we shall consider in some detail the analysis of the Brazilian Landless Peasant Movement (MST) on the development of agroenergy.

As capitalist logic has been at the origin of the destruction of nature and its irrational utilization of energy sources, one might have thought that the socialist countries, both in Europe and on the periphery, would have formed an avant-garde movement in the ecological struggle. It did not happen. In the middle of the nineteenth century, Karl Marx, who had inspired them, declared that capitalism destroyed the two sources of its own wealth, nature and labour. The ecological disasters that took place in several socialist countries, particularly the Soviet Union, were based on two main principles. The first was that it followed the scientist line of modernity, adopting the belief in linear progress, for which the control and exploitation of nature was necessary. The second was of a practical nature, both the desire to catch up with capitalism in order to increase the capacity of consumption by local populations and the mega-armaments imposed by the Cold War. This inevitably implied a predatory approach to nature. It explains why few precautions were taken in exploiting the mines, the diversion of rivers and the imprudent use of atomic energy. Such a model was all the more easily accepted because it was not linked to the accumulation of capital in private hands and was intended to serve the welfare of the public. To this should be added the authoritarian drift of a political system that was increasingly undemocratic and the economic planning that was insensitive to anything that could delay immediate economic benefit and its ultimate distribution in function of social needs. An intensive consumption of fossil energy was one of the consequences.

From the beginning of the twenty-first century there has been a tremendous development of a new conscience among world public opinion. Successive crises of the market have ended by warning citizens about the destructive relationship the neoliberal market has with nature and particularly in its use of energy resources, which was even more dire than its social consequences. Such growing awareness could be seen in two major developments. First, the protests against the world economic decision-making centres like the World Bank, the IMF, the G8, the WTO and the EC. The most striking event was in Seattle at the end of 1999, where there were demonstrations against the first meeting of the WTO, that had recently risen from the embers of the General Agreement on Tariffs and Trade (GATT). Participants joined in from a wide array of movements and organizations, who had never before demonstrated together: worker trade unions from North America, peasant movements from Latin America, indigenous peoples' movements, feminist and ecological movements, development NGOs, etc.: what Michaël Hardt and Toni Negri have inappropriately termed 'the multitude'. Everyone came together to protest against the decisions of a common enemy, of whom the various groups constituted the victims, and rather than protesting about this or that grievance, this movement of movements is coming together to challenge the system itself.

At the same time there developed a convergence among a collection of movements and organizations, thanks to several initiatives, such as Peoples' Power Twenty One (PP21) in Asia and the intergalactic meeting of the Zapatistas in Chiapas, Mexico in 1996. In January 1999, the Other Davos, having met in Zurich and then in Davos during the annual meeting of the World Economic Forum (WEF), brought together five large social movements from different sectors: the Landless Peasant Movement of Brazil, the worker trade unions of South Korea, the agricultural cooperatives of Burkina Faso, the women's movement of Quebec, the unemployed movement from France and a number of analysts, including Susan George, Riccardo Petrella, Samir Amin and François Chesnais.[2] This event marked the start of an important social phenomenon, that of the social forums at world, continental, national levels, as well as thematic ones. A new social conscience was being developed at the global level and networks of actors were reinforced and created in the aftermath of this new dynamic. The problem of ecology and that of the use of energy resources were very much present, even if it was only the dawn of a new process.

Then there was a series of events of a political nature: the UN Climate Change Conference at Kyoto in 2002, followed by that of Nairobi in 2007; the Intergovernmental Panel on Climate Change (IPCC) in Paris and Brussels in 2007; the European Summit in the same year, and the UN Climate Change Conference in Bali in December 2007 to follow up Kyoto; the experts' meeting on climate change at Poznan in 2008, etc. The rhythm of these events accelerated, becoming a key element in international negotiations. But the insane idea, typical of mercantile logic, of a 'right to pollute' made its appearance during the evolution of political consciousness. It proposed that countries emitting most of the greenhouse gases should come to an agreement with countries that were still 'clean' so that the latter did not adopt the model of destructive environmental exploitation and could thus dispose of their pollution quotas by selling them. This logic was pernicious, as it made it possible not to challenge the system, while mitigating some of its more negative effects. True, the worst attacks against the ecological equilibrium were to be reduced, but without affecting the mode of exploitation of natural resources that the logic of the system entails. So we are plunged once more into the question of sources of energy and their ecological effects.

Which brings us once again to Al Gore's film, *An Inconvenient Truth*. It is an excellent synthesis – as we have said – of the state of knowledge about the ecological destruction of the planet. It strikingly illustrates how the natural universe is deteriorating because of human activity. In the United States it reached a point where the Bush administration tried to minimize the gravity of the situation and certain officials did not hesitate to modify the conclusions of the scientists to support the political power's position. It was a question of conserving, at all costs, the American Way of Life and not touching the interests of the United States at the world level. The good conscience of the American people was thus reinforced in its belief in their country's model of economic development and its democratic management of the universe.

In contrast, Al Gore's film did show a certain courage, as it entered into direct contradiction with the American ethos. It showed that human practices were at the origin of ecological destruction and that if the whole of humanity acceded to the level of consumption of American citizens (and thus the damages inflicted on nature) the planet would be emptied of its animal and vegetable kingdoms and, finally, of life itself. It is also possible to imagine that the considerable efforts being made in space research by the United States are linked, among other things, to the idea that one day new spaces should be conquered and new astral bodies colonized so as to continue the existence of humanity and, first of all, of the American people.

The problem with the film by Al Gore, who is described by Corinne Lesnes[3] as a 'reasonable revolutionary', is that it leads to conclusions that are extremely misleading. Thus the solution to the ecological problem, of which the exhaustion of certain non-renewable energy sources is a part, is the responsibility of individuals. Clearly, this is not wrong and, as we repeat, it is desirable to make the message heard. But by appealing to individual consciences the author ignores or pretends to ignore the structural causes of the phenomenon. If the exploitation of natural resources, particularly in the field of energy, has contributed to such spectacular growth without taking social and ecological externalities into account, it is because it has participated decisively in the accumulation of capital. Without it, profits would not have exploded, fortunes would not have been built, shareholders would not have been able to assert their power, and a minority would not have been able to monopolize most of the benefits of development.

We are now waiting for another film that will show how the irrational exploitation of natural and energy resources is carried out, one that will reveal the names of the collective actors in this kind of activity. It will highlight the practices of the huge oil, mining and agrobusiness multinational corporations, as well as the compromises of state officials serving the multinationals or acting as intermediaries between foreign interests and the exploitation of their own countries. We shall then reach other conclusions. It is a whole system which is put into question and not only the behaviour of individuals.

The practices of corporations should also be studied to understand the delays in taking protection measures. Why have the oil lobbies held up research on and the production of non-polluting energies for so long? Why does their publicity now boast how 'green' their activities are? Why is there such a silence about the ecological and social costs in the production of certain energies, even those called 'bio'? The answer is clear enough, as will be illustrated later on. It is a question of preserving the sources of profit and not embarking on adventures that could prove expensive – which means that private interests must come before the collective good. All that enables, as we shall see, private capital to maintain its role as the engine of the economy. This is why a new film, *An Inconvenient Truth No. 2*, would be very opportune.

There is no doubt that the simple nationalization of energy resources is not enough to solve the problem. This has well been illustrated by the socialist countries and now by contemporary China. Two factors have to be combined and we shall speak more about them in our conclusions: there must another philosophy of our relationship with nature, and a democratic control of the common good of humanity – and thus of the energy issue.

The emergence of agroenergies cannot be understood except by situating them in such a general framework. They have been known about for a long time, and in a country like Brazil, already in 2006 there were 300,000 vehicles using 'green energy'. Europe seems to have discovered them and is making similar proposals. Everywhere people are celebrating the birth of a new saviour: the agroenergy that will make it possible to replace fossil energy. They say that CO_2 will no longer be poured into the atmosphere and in time the equilibrium of the climate will be restored. Certainly, the energy sector is not the only one at fault in making the climate so precarious, but it plays a key role, and thus an improvement in this field would mean a step forward, with eventually a hope for a definitive solution. This is the current discourse today and it seems to be accepted unanimously by the political world.

This is why the question of agroenergy should be looked into. Is it really a solution? If it is, what are the conditions? As it is an essential subject for the reproduction of life on earth and thus for the survival of humanity, it is worthwhile tackling the question in all its dimensions.

2
The Twin Crises: Energy and Climate

There are two big issues relating to energy use today. First, there is the widespread desire to use renewable resources, rather than those which will one day run out. Renewable resources include those perpetually in existence, like water, wind and sun, and those that are derived from plants which can be grown year-on-year such as wheat, maize, soya, sugar cane or oil palm (known as the African palm). Fossil energies are not renewable, even if some of them still have a fairly long life ahead of them (such as coal, as opposed to oil, gas and uranium).

A second concern is safeguarding the environment and the climate. The use of fossil fuels results in an enormous discharge of CO_2 and minute particles into the atmosphere, which is resulting in both global warming and pollution. Hence the desire to find alternatives, which would make it possible to create a better life for humanity as a whole and to preserve the integrity of the world's climate and ecosystems. The world is currently undergoing a double crisis that adds to the systemic upheaval of contemporary capitalism and, not unconnected to the latter, the energy crisis, which started with the oil shock, and the climate crisis which became visible at the beginning of the 1970s, although a general awareness of the issue did not develop before the outset of the twenty-first century.

THE ENERGY CRISIS AND NON-RENEWABLE ENERGY SOURCES

The energy crisis arises from the foreseeable end of a cycle, that of oil, gas and coal, which, furthermore, has considerably increased greenhouse emissions, a major cause of climate deterioration. Energy security is one of the main concerns of the principal economic centres of the planet and there would be no such security if there were to be an interruption in the oil supply or if the sources of oil were exhausted. For this conditions the possibility of growth, indispensable for the capitalist market economy and its development model. Following this logic, general world consumption has been predicted to increase by 60 per cent between 2002 and 2030, so it is necessary to find substitutes for fossil energy at all costs.[1] The demand for electricity, in particular, which was a little under 15,000 megawatts (MW) in 2000 will rise to 26,000 MW in 2025, which will require considerable effort.[2] According to the UN International Energy Agency (IEA), there will need to be 22 trillion dollars-worth of investment between now and 2030 in energy infrastructures to satisfy the increase in demand. Clearly, the stake is substantial.

Oil, gas, coal and uranium are the non-renewable energies and we shall look at them one by one. The first three are of fossil origin and they supply 80 per cent of world consumption. The IEA calculates that in 2030 the proportion will remain much the same, although coal could become more important. These three sources play a central role because they have a high energy yield. Interestingly, the first form of diesel was vegetable oil (similar to contemporary biodiesel) but it was soon replaced by oil which was more efficient.

In 2006, oil supplied 35 per cent of world energy, coal 23 per cent, and gas 21 per cent. At that time, these sources of energy were expected to last 40 years for oil, 60 years for gas and 200 years for coal.[3] It is not easy to find substitutes, at least for the medium term. Let us take just one example, agroenergy. It is estimated that this sector will represent only 2 per cent of consumption in 2012 and that it could rise to 7 per cent in 2030, which would require using the equivalent of all the agricultural land of Australia, New Zealand, South Korea and Japan. Even if all this cultivable land was concentrated on the production of energy, it would still only produce 1,400 million equivalent tonnes of oil. However, present needs are for 3,500 million and they are constantly increasing. There really is a crisis looming and the whole problem lies in finding out how to avert it: new energy sources, particularly non-renewables, savings in the various fields of consumption, another development model? This is only one aspect of the problem and the climate is inseparably linked to it. So now let us look more closely at the non-renewable sources, those of fossil origin like oil, gas and coal, and that of mineral origin, uranium.

Oil

'It was 40 years ago that the end of the oil era was predicted … in 40 years' time', a humorist has said. He was not wrong, because the discovery of new deposits and the use of increasingly advanced technology has made it possible to delay the deadline. But nothing can stop the end, even if it can be delayed a few more years. Whether peak oil (when oil extraction starts to diminish) is predicted for

2010 or 2020, or even whether it has already happened, as some say, is not important. In 2004, the IEA indicated that out of 48 producer countries, 35 were in decline, including Norway (–7 per cent), the UK (–10 per cent), Mexico, Oman, etc. While in 1950 the US was self-sufficient, the country reached its peak production in 1970, and in 2007 it had to import 75 per cent of its consumption. It is using a quarter of world oil production, while its known reserves are no more than 3 per cent.

If only to compensate all these reductions, the Organization of the Petroleum Exporting Countries (OPEC), whose members possess the most abundant reserves, would need to increase their production by 3 million barrels a day. But the demand is increasing all the time. It stood at 83 million barrels a day in 2005 and could reach 115 million barrels in 2015.[4] Nevertheless, proportionately oil has diminished in energy mass, from 50 per cent in 1975 to 36 per cent in 2006.[5] Oil is above all important for transport and heating. It provides 7 per cent of what is needed for world electricity production. It is true that the spectacular increase in its price has encouraged saving.

The main reserves are Saudi Arabia (264.3 billion barrels), Iran (137.5 billion), Iraq (115 billion) and Kuwait (101.5 billion). Russia has fewer than 100 billion barrels. Even if some believe that these forecasts are too optimistic, they give an idea of the order of magnitude. At all events, they make it possible to understand, without much further explanation, the geostrategy of the US in the Middle East. Iraq, for instance, has oil with a small amount of sulphur and its extraction cost is not more than two dollars a barrel. In 2007, with a production of 2 million barrels a day, 1.6 million were exported. It is therefore, as Mahomed-Ali Zainy of the Centre for Global Energy Studies of London has said, 'a petrol Eldorado in which the majors (the five largest oil companies) want their share'.[6] Thanks to the war, the oil company Chevron is now dominant there, replacing Total which was previously active in the country (ELF at that time). But an alliance between the two giants, combining field experience and strategic advantage, ended in 'an ideal marriage' according to Rula Hissani of the Energy Intelligence Group.[7]

Economically speaking, even if national companies mostly control the deposits, the international economic management of the sector is still dominated by the 'majors': Exxon, Shell, Chevron, BP and Total. In 2006, Exxon had a turnover of 450 billion dollars, more than the gross domestic product (GDP) of 180 out of 195 member states of the UN. However, according to Professor Patrick Brocoren, of the University of Mons, the production controlled by these companies has fallen by 5 per cent between 2001 and 2006.[8]

The spectacular increase in the price of oil during 2007 brought the price of a barrel to 100 dollars, but it evidently did not change the foreseeable decline of this source of energy. It was even partly one of the consequences. It both encouraged the pushing back of the frontiers of the exploitation areas (indigenous territories, natural reservations) and increased the competition between the consumers in the old industrial countries and those in the emerging countries. As for the increase in the price of oil – a function of its scarcity – this is already affecting the poorest, who are unable to afford the repercussions on the price of transport, food products and heating. Thus the logic of the reproduction of the system would mean strengthening the purchasing power of the 20 per cent better-off strata of the population. At the beginning of 2008, Exxon announced profits for 2007 as being at a record high in its whole history.

It should be added that the use of heavy hydrocarbons, asphalt sands and bituminous schist, which abound in the US and Canada, will hardly be a satisfactory solution because of the prohibitive cost of exploiting them, even if the spectacular increase in the price of oil has raised greater interest in them recently. Plus, there would be environmental consequences – in the Canadian province of Alberta, which is believed to contain the equivalent of 176 billion barrels, some 300,000 hectares of forest would need to be exploited, endangering some 160 million migratory birds, according to Jeff Wells at the Laboratory of Ornithology at Cornell University in the US.

Gas

Natural gas emits fewer greenhouse gases than oil, about one third less, according to the calculations. It is practically inoffensive in sulphur dioxide (SO_2). But, like oil, its reserves are limited. Natural gas represents 21 per cent of total energy consumed globally, and 20 per cent of the world's electricity is produced from gas, but it is expected to run out in a little over 50 years.[9] Another calculation, taking into account the whole of world energy consumption, comes to the conclusion that if gas alone were to meet global energy demand it could last for 18 years. These are the orders of magnitude. What is certain is that some deposits are declining. Those of the UK and of the US have declined, respectively, 10 and 28 years earlier than forecast.

In November 2007, during the Sixth Natural Gas Conference, which was held at Doha, the representative of Qatar, the main producer of liquefied gas, argued for a formula which was less volatile and better able to meet a demand that was constantly increasing. He proposed that the price of gas be delinked from that of oil. Also, investments are being made to recover methane from collieries. Thus Lorraine in France could provide 30 billion cubic metres, and in the Walloon area of Belgium reserves are calculated at 2 gigametres, the equivalent of the production of 10,000 wind turbines over 25 years.[10] At Lons-le-Saunier, in the French Jura, the European Gar Ltd, a branch of Australian Kimberley Oils, proposes to utilize methane. However, that is not going to delay the deadlines to any great degree.

Coal

It is commonly believed that there are far greater reserves of coal than either oil or gas, and that these can last 200 years based on current consumption. In fact, the reserves seem to be exaggerated: Energy Watch estimates that the peak will be towards 2025.

It is estimated that coal usage discharges 9,000 million tonnes of CO_2 into the atmosphere every year. In 2004 it amounted to 39 per cent in the world production of electricity (in China, 67 per cent). In the knowledge that the demand for electricity will increase by 60 per cent by 2030, there is a strong temptation to rely on coal to meet this demand. This is what the emerging countries are doing. China and India are to build 800 new coal plants between now and 2012, which will use five times the reductions forecast by the Kyoto Protocol for the planet as a whole, according to Fareed Zakaria.[11] China, in particular, has 118 billion tonnes of coal reserves, equivalent to 13 per cent of world reserves or half a century of its current usage by that country. Added to that terrifying fact is that in 2006 coal mines cost the lives of 6,000 workers in China. But it is not only the emerging countries that are thinking about coal because of the rapid increase in the price of petrol. The US foresees the construction of 80 new plants before 2012.

The amount of pollution through the thermal production of electricity based on coal was revealed unexpectedly on 15 August 2003 when the north-east of the US and the south-east of Canada experienced a breakdown in electricity supply which lasted 29 hours. The result was that there was a reduction of 90 per cent of sulphur dioxide in the atmosphere, of half of the ozone in the troposphere (the lower region of the atmosphere), and of 70 per cent of the particles attaching themselves to the clouds, and increasing visibility by 40 kilometres. Less electricity meant less coal combustion.

One solution proposed to remedy these disadvantages is to bury carbonic gas in the earth or the sea. The CO_2 emitted by the electricity power plants using fossil fuels (above all coal, but also oil and gas) would be captured and injected into the ground: this is called carbon capture and storage (CCS). Thus, as from 2012, Veolia in France proposes to deposit 200,000 tonnes of CO_2 into the Solin aquifer at Claye-Souilly in Seine et Marne, at a depth of 1,500 metres, while Total announces that, as from 2009, it will stock 75,000 tonnes of CO_2 at Lacq in the Atlantiques-Pyrenées. The Swedish company Vattenfall intends to do the same at Aolborg in Denmark in 2013. The European Union (EU) calculates that gas emissions could be reduced by 20 per cent, but this would entail an extra cost of some 6 billion euros, just for building the dozen 'demonstration' plants from now until 2015. The supporters of this solution maintain that this would give much faster results than nuclear fusion, for which 10 billion euros have been earmarked.[12] As for burying the carbonic gas in the seas, this risks contributing to the heating of the oceans, with a number of consequences which we shall talk about later.

Certainly, gasification and liquefaction of gas can improve performance when combining these processes with the burying of CO_2. China could stock 1 billion tonnes. Saudi Arabia has just set up a fund of 3 billion dollars for research in this field. The utilization of coal in electric power plants could, with these precautions, reduce an additional 15–20 per cent of greenhouse gases. However, there is another disadvantage in that these technologies use a lot of water, and for China this is not good news as it is chronically short of water. The American company Ashmore Energy, specializing in the gasification of coal, claims to have improved performances up to 90 per cent in pollution reduction. This is what the company has proposed to the government of Nicaragua. At all events, confronted by the scarcity of other energy sources, coal has once more come to the fore. In the *département* of Nièvre, for example, it is hoped to obtain 250 million tonnes.[13] But the massive use of coal will not help in the medium term to improve climatic effects – on the contrary.

Uranium

It was in 1942 in the US that the first reactor came into being, making it possible to transform heat into electricity. Nuclear power provides about 16 per cent of world energy production and France stands out in this respect, as in 2006 it produced 76 per cent of its needs, with 59 reactors. Across Europe as a whole, 164 reactors supply 28 per cent of electric current. In the UK, which has the oldest nuclear complex in Europe, the 14 plants in use supply 20 per cent of the electricity, but nine of them have to be closed down before 2015. In terms of immediate efficiency, nuclear energy has numerous advantages. Thus, 25 kilograms of uranium produce 1 gigawatt (GW) of electricity which would otherwise use, in a thermal plant, 2,700 tonnes of coal. Moreover, the 441 nuclear plants in the world avoid emissions of 3,000 million tons of CO_2.[14]

The increasing cost of oil has meant that the cry of the ecologists to halt nuclear energy is diminishing. Not only is France continuing to sell its reactors around the world, but the UK decided, at the beginning of 2008, to reconsider its policy, and so has Finland and Japan. Only Germany wonders what decisions to take in this field. But it should not be forgotten that uranium is not renewable and that its reserves are also diminishing. All its resources together would meet world energy requirements for just one and a half years, if it were the only energy source.

It is true that the technology has been evolving. In 2005 France launched the third-generation pressurized water reactor (PWR) and it is preparing a fourth generation reactor for mid-century, particularly plants using the rapid neutron capture process and 238 isotopes, the reserves of which are calculated to last thousands of years. This has been announced by the Commissariat à l'Energie Atomique (CEA). Also, fusion (two positively-charged nuclei) can one day take the place of fission (splitting the uranium nucleus in two). At the beginning of 2008 work began at Cadarache (Bouches de Rhone) on the International Thermonuclear Experimental Reactor (ITER) which, as already mentioned, has received an investment of 10 billion euros and is supported by 34 nations. Fusion has the advantage of not overheating (as in Three Mile Island and Chernobyl) and its waste material has low radioactivity, with a short lifespan. Experts are optimistic, but not everyone welcomes this technical challenge with enthusiasm. There are some people who, discouraged by the failure of Super-phoenix, describe it as a pharaonic project, which has little chance of succeeding.[15]

One of the problems hampering the development of the nuclear solution is that of its waste material. At the present time there are three formulas: reprocessing the used fuels, although this still leaves unburnt what is called 'hazardous waste'; underground or underwater storage, but this poses problems for future generations as it will be thousands of years before this material loses its radioactivity; or kept pending, which cannot last indefinitely.

Reactions against the dangers of storage are growing more and more frequent round the world. In the US, the federal government decided to store in Nevada, the spent fuels of the hundred or so nuclear plants supplying 20 per cent of the country's electricity supply as well as the waste resulting from military activities. The plan was to create 65 kilometres of tunnels at 300 metres below the subterranean water table, and to place the nuclear waste in 11,000 cylinders made of a combination of metals, which can be covered by a supplementary protection of titanium. The Department of Energy stated that this would not begin to rust for 80,000 years. The inhabitants of Nevada are afraid that water infiltrations through faults in the rocks would endanger the project and that the site was too close to the city of Las Vegas, which is in full expansion. The project, which was meant to start operations in 1998, was postponed until 2020 and it has already cost 11 billion dollars. Meanwhile, temporary sites are holding the waste material in 121 places and the producers are attacking the federal state in court because of the accumulated delays which have already cost them 300 million dollars.[16] In March 2009 the Obama administration stated that the site was no longer an option and proposed eliminating all funds in the federal budget except what was needed to answer enquiries from the Nuclear Regulatory Commission while the administration worked out a new strategy for nuclear waste disposal. However, in July 2009 the House of Representatives voted 388 to 30 not to defund the project in the 2010 budget.

Advocates of this source of energy draw a positive comparison with the ecological consequences of producing electricity through coal. With 25 kilograms of uranium one can produce 1 GW of electricity, which would require 2.7 million tonnes of coal and that would produce 3 billion tonnes of CO_2. We shall of course return to this aspect of the problem. There are some people, like Patrick Moore, president of Greenspirit Strategies in Canada, who think that the 441 atomic plants functioning round the world in 2006 prevent, as we have said, the emission of 3 billion tonnes of CO_2.

However, before closing this subject, we cannot ignore the risk of nuclear accidents, even if this is controlled to a large extent. Nor should we forget that in this field, the health and social conditions of producing the raw materials are often lamentable, if not criminal, especially in a country like Kazakhstan, and also in Africa. As for the negative effects on health of the operations in the plants themselves (for example leukaemia and other forms of cancer) they have been highlighted in numerous studies in Germany, the UK and France.[17] In the present situation it can only be said that nuclear energy is not a convincing solution, either for the long term or for the current ecological problems.

It must therefore be concluded that the crisis of non-renewable energies is real. At the present rhythm their usage will exhaust all the reserves in the world by the year 2100. Hence the interest in sustainable solutions and, in particular, for agroenergy. The key question is to know to what extent the challenge can be met. But before that we should also consider the incidence of energy use on global warming.

THE CLIMATE CRISIS OR GLOBAL WARMING

It is evident that the climate crisis is the result of human activity. The IPCC, which is responsible for establishing the bases of climate physical science, declared, in 2007, that this fact was 90 per cent proven. For a time it was believed that solar activity could, to a large extent, be responsible. However, what was called 'solar radiation' turned out to be two times less important than what was thought in 2001 and it will be ten times less important than the effects of greenhouse gases due to human activity.[18] Fuel consumption is a key factor: for example some 800 million cars in the world, which were listed at the beginning of the millennium, consume more than 50 per cent of the energy produced.[19] It is estimated that in Europe 22 per cent of CO_2 emissions are due to cars. However, as Anthony Giddens recently wrote, there is a paradox: 'since the dangers posed by global warming aren't tangible, immediate or visible in the course of day-to-day life, however awesome they appear, many will sit on their hands and do nothing of a concrete nature about them. Yet waiting until they become visible and acute before being stirred to serious action will, by definition, be too late.'[20]

The Role of Greenhouse Gases

The heating up of the earth is nothing new. Research on glacial layers has shown there was a long period of climatic warming, which lasted some 200,000 years, about 55 million years ago. It caused a series of great volcanic eruptions which increased the level of CO_2 in the atmosphere.[21] But, according to the same authors, the natural emissions remained stable for half a million years. During this period there was a compensatory effect, because plants absorbed an enormous quantity of CO_2 that had been discharged into the atmosphere by the decomposition of organic matter. The increase of greenhouse gas emissions began with the industrial revolution and it considerably increased during the neoliberal phase of capitalism, i.e. from the 1970s. This development model brought about, as we have seen, the exponential economic growth for a minority of the world's population, including the industrialized parts of the periphery. The importance given to exchange value, more than ever before thanks to liberalization policies, increased the mobility of capital, goods and services and the utilization of fossil energy.

Nitrogen trifluoride (NF) can now be included with carbonic gas and methane because, according to scientists at the Scripps Institution of Oceanography of the University of California at San Diego, this gas has a heating effect 17,000 times greater than that of CO_2 and has a life more than six times longer.[22]

The new information and communication technologies made it easier to separate production and consumption, but at the same time they increased the need for transport and hence the pollution of the atmosphere. Individual mobility was promoted and encouraged. Urbanization grew apace, particularly because of the rapid transformation of peasant agriculture into agrarian capitalism. This orientation is, in fact, one of the new frontiers for the accumulation of capital and it prioritizes monoculture, which by definition destroys biodiversity and is a voracious consumer of chemical products. This process has in fact been amplified by the production of agrofuels. And the social corollary has been the migration into towns. According to the International Urbanization Conference, which was held in Madrid in 2007, 66 per cent of the world population will live in urban areas by 2050. And it is known that cities are responsible for 70 per cent of greenhouse gases.

The power of financial capital enables it to fix the rules of the game of the world economy, giving priority to financial yields, not only to the detriment of productive capital, but ignoring the social and ecological consequences which are more than ever considered as externalities. In brief, the type of consumption that results from neoliberal logic demands increased exploitation of natural resources, particularly in the field of energy. Multiple pollution is the result and the climate gradually deteriorates. Economic progress has been advancing at the expense of the climate. It is calculated that a European produces an average of 10 tonnes of CO_2 a year and a North American 20 tonnes compared to a little over half a tonne by a Nigerian or about a quarter of a tonne on average by each inhabitant of Sudan.[23]

The greenhouse effect is the result of a complex natural mechanism. First the earth receives 50 per cent of the energy sent by the sun (the rest being absorbed by the atmosphere). Then the earth discharges this energy in the form of heat which in turn is partially absorbed by atmospheric gases. In the third stage the atmosphere releases some of this energy again towards the earth (which is what is called the greenhouse effect). When more greenhouse gas is emitted into the atmosphere from the earth's surface, the atmosphere absorbs more heat, some of which later comes back to the earth's surface, allowing the average temperature on the surface to rise. The greenhouse effect is not in itself negative. Without it the average temperature of the earth would be −18 °C, instead of 13–15 °C. It is thus a natural phenomenon through which the atmosphere keeps part of the solar energy and thus maintains a certain temperature on the planet. It becomes negative when the concentration of certain gases increases to the point at which the earth heats up more than is necessary for the life of living species, thus destroying the thermal equilibrium. These include CO_2 (which occurs naturally and is necessary for plant growth). But they are discharged in an excessive quantity by human activity. Methane (CH_4), the main element of natural gas, is produced by the fermentation of organic waste (particularly in rice fields under water, but also by domestic waste and the decomposition of cellulose in the stomachs of ruminants – animals that chew the cud such as cows, sheep and goats, whose numbers have escalated as demand for meat has increased). Then there are azote dioxide and nitrous oxide which result from forest fires, the combustion of fossil energies and the utilization of fertilizers. To these should be added the ozone in the lower levels of the atmosphere (the troposphere) coming from human activities that generate carbon monoxide and nitrogen dioxide.[24]

However, it is CO_2 which is the main agent of destruction. We emit 300 million tonnes into the atmosphere each year. It is responsible for 81 per cent of the emissions of industrialized countries, with 10 per cent coming from methane, 6 per cent from nitrous oxide (N_2O) and 3 per cent from other gases.[25] Even if the calculations vary, the proportions are similar, which is why CO_2 is given particular attention. In the 650,000 years of the climate's history that we know about from air balloons trapped in the glaciers of the Antarctic, the second half of the twentieth century alone exceeds by 130 per cent the massive carbon emissions observed during the course of the last five glacier cycles.[26] It is estimated that emissions into the atmosphere of CO_2 alone have increased by 130 per cent since 1750, and that of methane by 150 per cent, caused by the use of fossil fuels such as oil, gas and coal, and by the destruction of tropical forests and ecosystems that could counteract such effects.

Today, according to Pierre Friedlingstein, of the Laboratoire des Sciences du Climat et de l'Environnement in Paris, humankind is emitting eight times more CO_2 into the atmosphere than the billion tonnes that it produced in 1900, and three-quarters come from the consumption of fossil energies, the rest being due to deforestation.[27] In 2005, the World Resources Institute calculated that the sources of CO_2 emissions were divided as follows: 31 per cent industry, 19 per cent the housing sector, 14 per cent transport (of which 10 per cent on the roads, 2 per cent on the sea and 2 per cent in the air), 18 per cent agriculture and 18 per cent deforestation. As can be seen, the phenomenon is growing. In Belgium CO_2 emissions have tripled between 1961 and 2003.

According to the UN IEA, CO_2 emissions will increase by 57 per cent from now until 2030 if nothing is done to remedy the situation, and this will mean a rise of 3 °C in the temperature. From 1906 to 2005 the average global surface temperature increased by 0.74 °C. As for the average increase forecast for the twenty-first century, it ranges between 1.8 °C and 4 °C.[28]

While the use of fossil fuels is not the only source of climate pollution, it is still a very important contributor. It is, however, sometimes difficult to quote the amount in statistics. In Belgium, for example, a highly-industrialized country, it is calculated that greenhouse gases are due 40 per cent to industry, 22.8 per cent to the heating of buildings, 7.7 per cent to agriculture and almost 30 per cent to transport. In Europe as a whole, the figure for transport is 27 per cent. The United Nations Development Programme (UNDP) estimates that in developed countries, the contribution of the automobile to greenhouse gas is 30 per cent on average and this figure is continuing to increase.[29] In the US, according to Lester Brown, President of the Earth Policy Institute in Washington, the heating of buildings is responsible for 40 per cent of greenhouse gas emissions.

The increase of CO_2 in the atmosphere is above all due to the increase in individual mobility, which now engenders its own contradictions: congestion on the roads, slowing up of traffic, loss of working hours, waste of petrol, deterioration of road infrastructure, greenhouse gases (in France, road traffic increased by 43 per cent in the 15 years after 1990) and then, through the liberalization of a globalized economy which delocated and decentralized to the point where everything had to be transported. The widespread ownership of cars was part of a new orientation of the industrial economies, the principle of 'just in time', which releases stocks and conveys them to be assembled from different places. The environmental and social consequences are not taken into account in the financial calculations that prompt these shifts. Or, rather, they are socialized, i.e. paid by the collective (climatic damage, public works expenditure, health, etc.) or they are individualized, i.e. transferred to people, as if they represented the total of individual decisions and were therefore responsible (hours lost by shuttles, individual transport for going to work, etc.). Following such logic the external costs end up affecting business profit margins and hence the process of capital accumulation.

Another factor is therefore the explosion of transport in the neoliberal globalization of the economy. Between 1990 and 2003, world transport increased by 20 per cent. Growth was particularly strong for sea transport (26.2 per cent) and air transport (25.6 per cent). World maritime commerce rose from 2.5 billion tonnes in 1970 to 6.1 billion. The fleet of containers, which numbered 2,600 in 2000, grew to 3,500 in 2005 and 4,000 in 2008. In Europe air traffic exploded (more than 73 per cent between 1990 and 2003), mostly because of the low-cost companies. The forecast of the EC for 2012 is an increase of 112 per cent. In France the emissions of CO_2 by road transport multiplied 6.4 times between 1960 and 2000.[30] The generalized liberalization of trade promoted by the WTO encourages the globalization of commodities, whereby the law of competition profits from social inequalities and different ecological requirements (some call them comparative advantages).

All this has led to an absurd situation. Stéphane Lauer points out that in France, Chinese strawberries are very competitive in price compared with those from the (French) region of Périgord, but that their production and transport requires 20 times more oil. He also shows how the average distance covered by milk, fruit and plastic materials before they are converted into a pot of yoghurt is 9,000 kilometres. He cites Alain Morcheoine of the French Environment and Energy Management Agency (ADEME), who writes: 'Between a quarter and a half of the weight of a pair of jeans is emitted in CO_2 because of the delocalization of production.'[31] And there are many other examples that could be cited: the new season's fruit and vegetables from South Africa, fish from Lake Victoria, as well as flowers from Costa Rica, Colombia and Ecuador that land up in the supermarkets of Europe and the US.

Livestock is also one of the big culprits damaging the environment, according to the FAO in a report of 29 November 2006. From some aspects it is even more harmful than transport for it is responsible for 65 per cent of the emissions of hemi oxide azote which has a global warming potential 296 times greater than that of CO_2: it is attributable above all to manure. Moreover, livestock produces 37 per cent of methane emissions, as a result of the anaerobic fermentation of organic material during the digestive activities of ruminants (which is also produced in the flooded areas of rice cultivation). This gas is 23 times more harmful than CO_2. Each year total emissions attributable to livestock production amount to 70–75 million tonnes of greenhouse gases.[32]

To these figures must be added, also according to FAO, the fact that pastures take up 30 per cent of the land surface and that 33 per cent of arable land is used to produce livestock feed. In order to extend this still further, forests are being destroyed and 30 per cent of pasture land has been overgrazed, resulting in water pollution, the decline and erosion of soil, as well as a reduction in bacteria (micro-organisms) at the origin of the precipitation cycle (rain and snow) – what Professor David Sands of Montana University calls bio-precipitation in a study published by the journal *Science*.[33] Furthermore, this activity is one of the most damaging for water resources. It is expected that there will be a big increase in demand for meat between now and 2050, from 229 to 465 million tonnes, but always on an unequal basis: an Indian consumes, on average, 5 kilograms a year and an American, 123 kilograms.[34] In Belgium, the Flemish association Ethische Vegetarische Alternatief (EVA), in a public survey for the World Vegetarian Day of 2007, declared that the 285 million animals killed for consumption each year contribute to one-fifth of greenhouse gases.[35]

We have stressed the issue of agriculture, not to exonerate fossil fuels but rather to show that the irrationality of the model is the same in different fields. The objective is first of all to satisfy the consumption of the 20 per cent of the world's richest, without worrying about the collective cost that this entails, as long as it does not affect the profits of capital. Agricultural production, and essentially its productivist extension, is carried out along the same lines as we found in agrofuels: monoculture, agrarian capitalism, transnational corporations monopolizing inputs and marketing, the ill-considered promotion of genetically modified organisms (GMOs) and the destruction of peasant agriculture.

All this affects climatic change and brings with it a series of consequences. It has already been noted that the number of storms in the world has quadrupled between 2004 and 2007. Global warming is perceptible. The experts at the IPCC calculate that if the average global surface temperature increases by more than 2 °C there will be extremely serious consequences and, to avoid it, global greenhouse gas emissions must be reduced by 50 per cent between now and 2050. If this is not done, says the American climatologist James Hansen, Director of the Goddard Institute for Space Studies, in the *Annals* of the National Academy of Sciences of 29 June 2006, 'we shall have a planet that is different from the one we know'. This scholar has arrived at the conclusion that present temperatures are at the top of the range of those that have prevailed since the beginning of the Holocene period, 12,000 years ago.[36] For his part, Lester Brown is very radical in his conclusions. He believes that the IPCC is two years behind in its calculations and that greenhouse gases must be reduced 80 per cent by 2020 to resolve the problem of global warming.

Diminishing Carbon Sinks

It is not only the production of CO_2 that affects the climate. There is also the question of the capacity of absorption, what are known as the carbon sinks – the seas and the wooded land. Since 2002, according to the *Annals* of the National Academy of Sciences in 2007, the rate of CO_2 emissions has increased by 35 per cent, of which half is due to heavy use of fossil fuels and half to the decline in the absorption capacity of forests and oceans. While the forests are shrinking fast, the oceans are becoming increasingly polluted. These two carbon sinks – according to the same source – absorb half of the greenhouse gases emitted. Now the levels of CO_2 in the atmosphere are the highest for 600,000 years and perhaps even for the last 20 million years.[37]

Forests

There are different kinds of forests, those called natural or primary ones (which include tropical forests), and those that have been planted either for economic reasons (for exploitation) or for ecological reasons (for protection of urban areas). There are three major zones of tropical forest, the Amazon in South America, the rainforest belt of Central Africa and the tropical forests of South-East

Asia, largely located in Indonesia and Malaysia, as well as certain less important regions like Central America and Papua New Guinea. It is calculated by the International Tropical Timber Organization, a group formed by 59 countries, that the area covered by forests is 816 million hectares, although the FAO, with somewhat different criteria, estimates that there are 1,200 million hectares of forest land.

Natural forests are called carbon sinks: the trees absorb CO_2 to achieve growth. This makes it possible to recycle the air and thus to halt climate warming. All the forests in the world absorb between 3 and 4 gigatonnes (billions of tonnes) (Gt) of greenhouse gases each year.[38] All natural forest, of which the major part is tropical, absorbs 2 Gt of carbon each year, which represents a quarter of the world human production of CO_2.[39]

Each year 6 million hectares of primary forest are felled. Only 4.4 per cent of tropical forest is managed to ensure its conservation. The FAO also affirms that 3 per cent of forest land disappeared in the last 15 years, while the UNDP states that between 2000 and 2005, 73,000 square kilometres of forest have been destroyed, equal to the territory of Chile.[40] There are various reasons for the destruction: commercial exploitation of timber, encroachment by pasturage or other crops that are more profitable than exploiting the forest,[41] and the use of wood for fuel by local populations.

It is not difficult to give concrete examples. The Amazonian forest extends over 6,762 square kilometres, involving five key South American states: Brazil with 68 per cent of the area, Ecuador with 7 per cent (but which constitutes 41 per cent of its territory), Bolivia, Colombia and Venezuela. It contains more than 1,000 rivers, 20 per cent of the fresh water of the planet and it is the home of 80,000 plant species and 2,000 kinds of fish. It plays an important role in balancing the production of oxygen and the fixation of CO_2. According to Steven C. Wosfy, it absorbs 2.8 kilograms of carbon per hectare, per hour, while the breathing of trees produces 1 kilogram of CO_2 per hectare, per hour (Wosfy, 1998). Twenty per cent of tropical forest in Brazilian Amazonia disappeared between 1960 and 2005: the same as for the whole period of the country's history since the arrival of the Portuguese, 450 years earlier. Greenpeace gave a vivid picture of this destruction, describing it like the disappearance of a football stadium every two seconds. From a climatic viewpoint, such deterioration means the equivalent of an additional 1.5 billion tonnes of CO_2 each year. The main causes are the extension of agriculture and uncontrolled logging. In Brazil alone, more than 3,000 lorries illegally transport timber from the Amazon each day.

At the current rhythm of deforestation, with the help of climate change, the Amazonian forest will be destroyed before the 2040s.[42] The Ministry of the Environment published a report in 2007 stating that in 2100 the rise in temperature could reach 8 °C, which would reduce the whole Amazonian region to savannah land.

Among the leading causes of this destruction is the extension of monoculture to produce food for humans and animals and, increasingly, for agrofuels. As at 2004, 1.2 million hectares had been converted to soya, according to Greenpeace. Other causes include the plantations of eucalyptus for paper production and charcoal and also cattle ranching. There are currently 60 million head of cattle in Brazil, and the export value to Europe alone comes to 3.5 billion euros. Much of the destroyed forest is burnt, a process that constitutes 75 per cent of the pollution in Brazil. The consequences are serious indeed. The Amazonian forest is pluvial and the humidity it produces provides half of the rainfall in the region. Deforestation thus increases the drought which in turn encourages forest fires, diminishes the level of rivers and renders vulnerable the trees that have only a shallow layer of earth for their roots. All this has an impact on the climate so that a veritable vicious circle is set up. Brazil now is the fourth largest producer of greenhouse gases in the world. One of the factors, besides the extension of livestock production, is the uncontrolled network of roads (170,000 kilometres) to exploit the forests and rare timber. In 2007 the forest had diminished by 3.8 per cent.

It cannot be said that the authorities are indifferent to all this. Between 1993 and 2006, deforestation was reduced by 52 per cent. Numerous measures have been adopted to stem the process of destruction, but control has often been ineffectual. In 2008 the Brazilian government launched the Amazon Fund to Combat Deforestation which aims at collecting donations, of which 80 per cent will be destined for the Amazon and 20 per cent for other regions, including those outside Brazil. The aim is to raise 21 billion dollars between now and 2021 from governments and the private sector. Norway has been the first to contribute, with 100 million dollars. This initiative demonstrates the ambiguity of international interest in Amazonia. President Lula of Brazil expressed it clearly on this occasion, referring to 'the rich countries who speak as though they were the owners of Amazonia'.

However, this affirmation of national sovereignty contradicts other internal measures recently taken by the government. In July 2008 in Brazil, a law confirmed the Provisional Measure No. 422, to extend rural concessions from 500 to 1,500 hectares in the Amazonian territory, without any invitation for tenders and with the possibility of deforesting 20 per cent of the land. This decision provoked a strong reaction from Marina Silva, former Minister for the Environment.[43]

Sometimes the political powers are both judge and protagonist. At Mato Grosso, the governor himself is the biggest producer of soya in the country. Brazilians are very sensitive when one brings up the subject, as it infringes their national sovereignty. It is understandable, seeing some plans from the United States proposing to make Amazonia an international zone for the protection of the climate. In 2007, a geography textbook, used particularly in secondary schools in California, showed the extension of this zone covering seven countries in the region and pointed out that they were incapable of preserving this heritage of all humanity and that others should take on the task. One has only to recall the Monroe Doctrine with its 'America for the Americans' (of the North) to be suspicious.[44] But the Brazilians have the same reactions in other forums. The proposal to create, within the UN framework, an International Environment Organization, does not please President Lula. 'The rich countries are crafty,' he said on this occasion, 'they decree norms against deforestation after they have destroyed their own forests.' The Brazilian Institute for Agronomic Research (EMBRAPA) revealed that Europe had only conserved 0.3 per cent of its primary forests, as against 69 per cent that had been conserved in Brazil. But it cannot be excluded that the agro-export model that the Brazilian political economy has always played a negative role in this field. Their fear is that protective measures in the silvicultural zones will be introduced by the WTO, thus making obstacles for agro-exports. This also probably has something to do with the official reaction of the Brazilian authorities. The lobby of the 'ruralists' is pressing for a reduction of the forestry code.

On the rest of the Latin American continent there are similar situations too, but on a smaller scale. In Argentina, 300,000 hectares of wood are destroyed each year because of the extension of agricultural land, particularly for the monoculture of soya and eucalyptus. In Central America it was above all cattle raising at first and then cotton (during the Korean war) that caused deforestation, not to mention the search for precious timber. In less than half a century the region has lost three-quarters of its primary forests. In Colombia, a region like Chocó, which had exceptional biodiversity, has been almost completely stripped of its woodlands by the owners of extensive livestock ranches and the palm oil companies. The peasants and the indigenous communities of African origin have been brutally expelled from their land, as we shall see later.

In the Congo, forests spread over 1.5 million square kilometres. They have been given a certain respite by years of warfare but since 2002 concessions to foreign enterprises have multiplied. The forestry code, published that same year with the support of the World Bank, has a distinctly mercantile vision of forest management. Numerous enormous contracts have been the object of revision by the Congolese government. But the damage has been done. Whole regions have been devastated. The big timber companies came in with immense machines in order to take the best trees. The environment has been ravaged, the reforestation clauses ignored and the local population alienated, having lost a large part of their means of subsistence. Controls have been non-existent. According to Greenpeace, 21 million hectares have been exploited by these timber companies.

Indonesia and Malaysia have lost more than 80 per cent of their original forests, essentially to the African palm. One of the methods used to free land from forest has been by setting fire to it. In Indonesia, during 1997 and 1998, more than 3 million hectares were burnt, to the point that the country became the third largest emitter of CO_2 into the atmosphere, after the US and China, with a discharge of 2.5 Gt of carbon. On top of this destructive policy, global warming also reduces the humidity in the undergrowth, making them prone to fires. This does not happen only in South-East Asia.[45] Gradually we are losing forests as carbon sinks.

Oceans

The oceans absorb about 40 per cent of CO_2, which makes them a particularly important element in the climate equilibrium. But their regulatory function is diminishing because they are slowly becoming saturated by CO_2. The problem is the temperature of the water. The solubility of CO_2 in water increased with the lowering of the temperature. Oceans are carbon sinks when their water is cold but they become sources of carbon emissions when they become warm.[46] They are also under considerable assault. For example, 87 per cent of the waste water in Latin America is discharged without treatment into the rivers and seas[47] while the Mediterranean is on the verge of asphyxiation through pollution. To these should be added the dangers of the oil tankers that cross the oceans and the ecological catastrophes that have occurred in the US, France and Spain. Furthermore, each year 1.9 billion tonnes, 62 per cent of world oil production, is transported by sea,[48] thus adding a growing source of contamination. The phenomenon of the 'dead seas' has also increased because of the nitrogenous fertilizers used by monocultures which are carried down by the rivers to the sea.

According to the conclusions of the Third Debate on the Biology of Conservation, held in Majorca in 2007 under the auspices of the University of the Balearics, marine ecosystems deteriorate ten times quicker than tropical forests. The causes are the excess of toxic products and the contribution of nitrogen, which is responsible for hypoxia (lack of oxygen) that kills marine flora and fauna. From 5 per cent to 9 per cent of the coral reefs are disappearing each year and 2 per cent of the mangroves (a thick plant growth that protects shores in the hot regions).

Rising Temperatures

Changes in temperature are easily measured. Over the last 100 years, the average temperature of the planet increased by 0.7 °C. In the US, the year 2002 saw an increase of more than 0.71 °C – more than the average for the whole twentieth century. In Cuba, eleven of the twelve years between 1995 and 2006 were the hottest since 1850. According to the World Meteorological Organization (WMO), 2006 was the sixth hottest year since statistics first started to be collected – 0.42 °C higher than the base temperature (between 1961 and 1990), more so in the northern hemisphere (0.58 °C) than in the southern hemisphere (0.26 °C). Eleven of the last twelve years have been the hottest in the world for 150 years, according to the report of the IPCC, the relative cold of 2008 being due to the Niña phenomenon.

The British meteorological agency has announced that 2006 was the hottest for the last 347 years. In Belgium and the Netherlands, 2007 had the highest temperatures since meteorological records began at the end of the eighteenth century. As for the German Institute for Research on Fisheries, it concluded that the North Sea is going through its longest reheating phase since 1873, when measurements were first taken. Researchers like Philippe Bousquet, of the French Laboratory of the Climate and Environment Sciences, estimate that the recent increase of human-related greenhouse gas emissions is above all due to the growth of industrial activities in Asia.[49]

This is the case, among others, for China, whose annual growth rates are above 10 per cent and it is blamed for the damages being wreaked on the climate. According to the UN IEA the country will become the biggest consumer of energy in the world by 2010. The McKinsey Global Institute calculates that between 2003 and 2020 the number of vehicles in China will rise from 26 million to 120 million, that the area covered by human habitat will increase by more than 50 per cent and that the average growth in the demand for energy will be 4.4 per cent a year. In 2007 China approached the US in its level of greenhouse gas emissions and, according to the IEA, it will have very largely overtaken it by 2030. However, we should not forget that its population is more than four times greater than that of the US and that the most important sources of global warming have for decades been – and continue to be – the industrialized countries of the North.

Role of Chlorofluorocarbons

Apart from greenhouse gases, there are other sources of atmospheric pollution. Above all there are the chlorofluorocarbons (CFCs) which were used in aerosols and refrigerators and were responsible for degradation of the ozone (O_3) in the stratosphere (high atmosphere). At an altitude of between 10 and 14 kilometres, ozone absorbs most of the very short ultraviolet sun rays (UV-B) which are harmful for living creatures. It is therefore highly protective and its reduction is a serious danger, together with the attacks against superficial ozone in the troposphere. This trend has been increased by contaminants such as carbon monoxide and azote, which are formed by the incomplete combustion of certain fuels. Large quantities of them are spread into the atmosphere and together they either prevent the sun rays returning back to space, which gradually increases the average temperature of the planet, or they destroy the ultraviolet ray protection zones. The former is what is known as 'planetary darkness', which apparently stimulates an effect that is contrary to that of CO_2.

We should remember that the ozone layer is vital, otherwise the ultraviolet rays of the sun would kill off all life on earth. It is beneficial in the high atmosphere (but harmful in the troposphere, or lower atmosphere). When ultraviolet activity is too strong because the protection of ozone in the high atmosphere has been reduced, it is, for example, the cause of skin cancer. In 2006, at the end of the winter in the southern hemisphere, the ozone hole in the high atmosphere extended over an area of 3–4 million square kilometres, which is partially explained by the heating up of the Antarctic troposphere, because of the concentration of greenhouse gases. This causes a cooling of the higher layers which in turn affect the impoverishment of the ozone layer. As for the CFCs they also destroyed the ozone in the stratosphere (high atmosphere). They discharged micro-particles into the atmosphere from the use of aerosols, refrigerators, air-conditioners[50] and the combustion of fossil energies. Through the effect of the sun, the chlorine of CFCs is liberated and destroys the molecules of the ozone.

There is, however, one cheerful piece of news among all these worrying facts: the rate of ozone in the stratosphere has ceased to diminish because of the application of the Montreal Protocol of 1987. This was announced in a joint report of the WMO and the UNEP dated 18 August 2006. However, there will not be a return to normal before 2050 for most of the world, and before 2060–70 for the Antarctic. As for acid rain, which resulted from the same kind of problem, the situation has also improved.

Combating this phenomenon was much easier than the fight against CO_2. It was directly visible and therefore more easily perceived than carbon dioxide. It was enough to replace the gas used in the aerosols, refrigerators and air-conditioners and to burn fossil energy more cleanly (by catalysers) to get quite rapid results. The hydrofluorocarbons (HFCs) which replaced the CFCs are eight times less harmful and we shall gradually be moving towards using hydrochlorofluorocarbons (HCFCs) which do not destroy the ozone layer at all. An international agreement was reached at Vienna in 1986, and then in Montreal in 1987 and 20 years later, the results are evident, thanks largely to the replacement materials and the progress made in combustion engines.

In September 2007, again at Montreal, 200 countries were even in agreement to accelerate by ten years the elimination of substances harmful for the ozone layer. UNEP Director Achim Steiner considered this a 'vital sign' for the Bali conference. The phenomenon is thus on the decline. The success in this field shows that it is possible to take action where there is the political will to do so.

Micro-particles emitted by aerosols also have an effect on stratospheric ozone by producing a membrane of pollution which fixes itself to the clouds and prevents the light of the sun from reaching earth. The light fades, bringing about what is called planetary darkness. This has been observed, not only in various places in Europe, but also in the US, Australia, the Maldive Islands and Israel. In Australia, for example, the rate of evaporation of lakes has been going down over the last century. It is a paradox because elsewhere the earth is heating up. In Israel, there has been 22 per cent less sunshine in 20 years. According to Beata Liepert, a German researcher, sunshine has diminished between 1950 and the beginning of 1990, by 10 per cent in the US, by 16 per cent in certain areas of the UK and by 30 per cent in Russia.

In the US an unintended experiment took place in September 2001. After the terrorist attacks of 9/11, all air traffic was suspended for three days. As a result there was an increase in temperature of more than 1 °C, thanks to the absence of the micro-particles discharged by planes, which contribute to the darkness phenomenon. This has to some extent masked the heating-up process. According to Peter Cox, an American researcher, eliminating planetary darkness could cause global temperatures to rise by more than 10 °C between now and the end of the century. Hardly something to be happy about!

Effects of Climate Change

Calculations about the foreseeable effects brought about by climate change that we have described so far depend on two factors: one being the quality of the measurements of these transformations, and the other, the projection into the future of the observations of what has been happening these last few years. It cannot be denied that there is a degree of uncertainty. Some use this as an argument, as we shall see later, to cast doubt on the alarmist announcements about climate change. Still others – and they include the majority of scientists – are constantly refining their projections, acknowledging that they are not absolute truths, but probabilities that are likely enough to be taken seriously by preventive action. 'If most of the damage will take place in the future, prevention must start now', says Jean Pascal van Ypersele, professor at the Catholic University of Louvain and vice-president of the IPCC.[51] The same reasoning can be applied to an analysis of the consequences of global warming which we shall now tackle. The main thrusts will be indicated, illustrated by examples and with an emphasis on the logic that unites them. These observations and reflections can be summarized and considered from three major viewpoints: the environment, the economy and the socio-political consequences.

On the Global Environment

The first ecological effect concerns biodiversity, by which I mean not only the number of living species, but the extent of the gene pool and the resilience of ecosystems. Of course, biodiversity is also in danger from other factors apart from climatic change, such as pollution from industrial activities and the reduction of natural habitat caused by monoculture and urbanization, as has been pointed out by Etienne Brouquet of the Belgian Platform on Biodiversity. The IPCC estimates that if the temperature increases by 2.5 °C between now and 2050, 20–30 per cent of animal and plant species will disappear. According to the International Union for the Conservation of Nature (IUCN), which draws up a list of endangered species, out of 41,415 species of vertebrates listed in 2007, 16,308 are threatened by extinction – 200 more than in 2006. One mammal out of four and one bird out of eight. These figures were confirmed by the IUCN at its meeting in Barcelona in October 2008. Examples abound, from the great apes of Borneo to the polar bears, seals and penguins in the icy regions, as well as the dolphins of the Yangtze and oysters all over the world. In 2008, 50 per cent of the world's species of primates were endangered.[52]

The International Fund for Animal Welfare (IFAW) believes that in the gulf of Saint Lawrence in Canada, all the young seals drowned when the ice flows melted. In 2007, the US Federal Fish and Wildlife Service (FWS) estimated that 3,000–4,000 walruses alone had perished on the Russian shore of the Chukchi Sea. Not only have the ice floes melted but the plankton tends to disappear with global heating. As for the penguins of the Antarctic, their numbers have been reduced by 70 per cent in 30 years because of the disappearance of the krill, the minute crustaceans they feed on, which are also affected by the diminishing plankton that feeds them. All this is caused by the serious change in their polar habitat.

There is a mushroom that is proliferating because of global warming and is attacking the frogs of Central and South America – 67 per cent of the Harlequin Monteverde frogs have vanished.[53] It is also very significant that to remedy the disappearance of plant species, action has been taken to bring together in a grotto cut into a mountain at Longyearbyen (Swalbard) in the Spitzbergen islands, 4.8 million seeds at a temperature of –18 °C at 130 metres below sea level. The project is called Noah's Ark and it is financed jointly by the Norwegian government, the Bill and Melinda Gates Foundation and several seed companies like Monsanto. Already 268,000 samples of cereals have been gathered: the aim is to collect samples of 250,000 known plants and preserve them as a resource for future generations.

The heating up of the seas has also encouraged a veritable invasion of jellyfish, which are very harmful for other aquatic species, particularly in the Black Sea, the Baltic Sea and the Mediterranean. A conference of experts, held in Madrid under the auspices of the European Union (EU), stated in January 2008 that more than 10,000 exotic species was endangering European biodiversity (Daisie Programme, Delivering Alien Invasive Species Inventories for Europe).

In South Asia, climate change has upset the pattern of monsoons and in 2008 India experienced the worst floods in recent history. Peasants are being obliged to adapt their cultivation to the new conditions. For all these reasons Bruno David, Director of the Biochemistry Laboratory at the University of Burgundy in Dijon, posed the question: 'The biosphere has not been killed, but does it still have the same capacities to regenerate?'[54]

The tropical zones are in special danger. The coral reefs – 'real tropical forests of the sea' as Annelise Hagan of Cambridge University describes them – are particularly menaced.[55] At the same time, desertification is accelerating in Africa and will also be much affected. But other areas are not exempt from peril. In Europe there may be more heatwaves and droughts in the Mediterranean area. In the US, Hurricane Katrina and the forest fires in California are clearly due to climate change. However, according to some calculations, Scandinavia, Russia and Canada may benefit from a milder climate. But the cyclonic regions of the Caribbean and East Asia are likely to suffer climate catastrophes increasingly often and more violently. In 2008 three cyclones ravaged Cuba, causing almost 10 billion dollars-worth of damages and claiming hundreds of victims in Haiti. The El Niño/La Niña phenomena are affecting the coastal regions of the Pacific, such as Ecuador and Peru, more frequently, as well as the other side of the Pacific in the Philippines, through increased rainfall and flooding. Taking a more long-term view, researchers at the University of Wisconsin estimate that 48 per cent of the planet could have a different climate in 2100 than they have today, if changes are not made.

Special attention should be paid to the question of water. Oceans cover 72 per cent of the planet and water is an essential element for the survival of all living things. Therefore the consequences of climate change are very important in this field. Glaciers are melting across the world. The IPCC has announced that the small glaciers in Europe will have disappeared by 2050. In March 2008 the UNEP declared that the average rate at which glaciers are melting has doubled between the years 2004–05 and 2005–06, as calculated by the World Glacier Monitoring Service (WGMS). They have lost 11.5 metres of thickness since 1980 and Achim Steiner, UNEP Director, who is also a UN Assistant Secretary-General, declared that this could affect millions, if not billions of people in the world because of the repercussions on the supplies of water, industry and energy production.[56]

In the Himalayas, according to a UNEP report,[57] the massif is heating up between 0.15 °C and 0.6 °C each decade and the glaciers are shortening by 10–60 metres each year. In 24 years the 46,928 glaciers within China (including Tibet) have been reduced in volume by an average of 5.5 per cent. The process is leading to the formation of pockets of melted snow that accumulate in the lakes. In the Himalayas alone, according to the same report, some 2,000 lakes are dangerously full and risk causing huge 'mountain tsunamis' (Glacial Lake Outburst Flood – GLOF).

In the longer term the great rivers of the Himalayan region, fed by the glaciers, could be short of water: in India alone they supply 50–80 per cent of the main water courses. This could constitute a challenge for another important function. According to Professor Valier of the University of Nancy, the sediment brought down from the erosion of the Himalayan range, drained by the rivers, will add to the organic carbon of vegetable detritus and will end up in the Indian Ocean. This prevents the oxidization of organic matter in CO_2. The severe erosion of the land creates the displacement of over a billion tonnes of earth, which causes such a rapid sedimentation rate that the time of exposure to oxygen is limited, thus diminishing also the CO_2 in the atmosphere, which obviously makes the climate colder.[58] If this process slows down it will necessarily impact on climate change. This region covers eight countries, including some of the most densely populated in the world, with a total of 2.4 billion inhabitants. In contrast, in Latin America only 30 per cent of organic carbon transported by the Amazon river ends up in the ocean, but there are other factors. Thus, in the Andes, the same phenomenon of the melting of glaciers has already reduced the water in rivers, limiting the possibilities of irrigation and the supply of drinking water. This is particularly the case for Quito, the capital city of Ecuador: its nearby volcano Cotopaxi has lost 23 per cent of its ice mass between 1993 and 2005 as a result of an increase in average surface temperature of 0.8 °C over 44 years.

The melting of the ice in the Arctic, a region of 2.6 million square kilometres, is also accelerating. According to a report of the Geophysical Research Center, dated 2006, at this pace ice floes will have disappeared by 2040. In 2007 Michael Wallace of the University of Washington wrote that we have already reached a point of non-return.[59] In summer 2007, the Arctic ice sheet was reduced in size by an area equivalent to ten times the size of the

UK compared to previous summers.[60] Furthermore, permafrost in Alaska and Siberia is disappearing with the consequent emission of high levels of methane. On 5 April 2009, 40 kilometres of the ice bridge linking the Wilkins Shelf to the Antarctic broke up, showing that the process is accelerating. In the Arctic the ice floes are 42 per cent thinner than in 1958.[61]

It is hard to believe, but the melting of sea ice has only a marginal effect on the level of the oceans, contrary to the loss of land ice which produces huge quantities of water that run into the seas. Greenland is now losing 100 billion tonnes of ice each year.[62] In the Frozen North and the Antarctic many of the ice floes are already in the sea; their melting does not add much to the volume of water in the oceans, but the damage to the ice floes destroys the natural habitat of many animals.[63] However, it is above all the permafrost, that permanent layer of snow covering the tundra, that is disappearing. It has a reflective effect which, on being reduced, also contributes to the warming of the planet. But the climate changes in the Arctic also have another unforeseen effect: a new sea-way is being forged between the Atlantic and Pacific oceans and natural resources, particularly energy sources, are becoming accessible, stimulating the appetites of the bordering countries. In 2007 this came to international attention when a Russian flag was planted under water at the site of the North Pole.

In the Antarctic, it is the peninsula that sticks out into the sea and the oceans that surround the continent that are heating up, while the interior is becoming colder because of the winds that thicken the ice. This continent is very important for the climate: it extends over 14 million square kilometres and contains 70 per cent of the reserves of the world's fresh water. An international treaty declared it a demilitarized zone in 1959 and the Madrid Protocol of 1991 forbids all mining extraction there until 2041. The International Polar Year (2007–08) will no doubt be giving us more knowledge about what is happening and the exact state of its transformation. The Antarctic Ocean acts like a biological pump, absorbing CO_2 which is sucked down into the depths of the sea (3,000–4,000 metres). This process is the result of the photosynthesis action of plankton: the more intense the cold, the greater the absorption. But over the last 50 years the temperature in the Antarctic Ocean has increased by 2.5 °C. The icy season has been reduced by two weeks over 20 years. The floes are getting smaller each year and the ice cap is becoming smaller, while the sea level is rising at a more rapid rate than at any time in the last 5,000 years.

The presence of greenhouse gases (CO_2, sulphur oxide, methane) increases the acidity of the water which harms the plankton. Hence the idea of Alfred Wegener, a German researcher, to inject iron sulphate into the oceans saturated with CO_2. This would make it possible to capture 15 per cent of carbon emissions, thanks to the development of plankton. A further problem is the increasingly high level of ultraviolet rays because of the holes in the ozone, which disturbs the photosynthesis process. The Antarctic's capacities to regulate its environment are dwindling.

So we should take a closer look at the oceans themselves. There are several developments. First, with the melting of the glaciers their volume is increasing, which risks having a serious effect on coastal areas and low-lying islands. In one century the North Sea has risen 17 centimetres. In its 2007 report the IPCC estimated that the sea level was going to rise 18–59 centimetres during the twenty-first century. Bangladesh could lose 17.5 per cent of its territory if the sea level rises by 1 metre,[64] but a country like the Netherlands will not be spared either and the Dutch government has decided to allocate 1.9 billion euros to raise the dykes. The Pacific islands are directly threatened and they raised their voices at the Bali conference, asking – in vain – for funds to help them prevent such a catastrophe. A study by the Catholic University of Louvain calculated that if the sea level rose 1 metre the city of Antwerp would be on the coast. If it reached 8 metres, one-tenth of Belgian territory would be covered by water, and if it attained 15 metres, Brussels would be a sea port, rather than being linked by a canal as at present. Many of the world's great cities would be affected, including New Orleans, Mumbai, Jakarta, Lagos, Amsterdam, New York and London.

This is happening because of global warming which produces an expansion of water and is complemented by the melting of glaciers. Hence the changes in the circulation systems within the oceans themselves, which stock heat and redistribute it through currents, from the tropics to the poles. This caused, for example in the case of the Indian Ocean, a change in the seasonal rains in Ethiopia, which was responsible for the droughts that have been taking place in that country since 1996, according to a report by the Royal Society published in London in 2005. According to Thom Hartmann, global warming could also provoke changes in the Great Conveyor Belt, a cold and salty underwater current from the North Atlantic which, off the Greenland coast, plunges into deep water until the Cape of Good Hope, where it mixes with the waters of the Pacific Ocean. This current, which was discovered recently, has a determining effect on the temperatures of Europe and North America. If it changed, these regions could undergo a contrary effect to global warming: it could promote a new ice age.[65] In the far future, according to certain researchers, the course of the Gulf Stream could also be in danger.

It is therefore the capacity of CO_2 absorption that is at stake for the oceans. The increase in emissions together with the pollution of the seas by the discharge of waste water and toxic products as well as by the density of maritime traffic, reduces its capacity of absorption. All this aggravates the climate crisis as CO_2 is less and less absorbed by its natural regulators.

What is particularly disquieting about water, apart from the floods which will be the fate of the coastal regions and those affected by typhoons and cyclones, as well as the increasing levels of salt in certain soils, is the lack of fresh water. This risks, as from 2080, affecting 3.2 billion people, as was stated by the IPCC experts at their conference in Brussels in 2007. It is already a very real problem in the Middle East and in the Sahel.[66] According to Marc Gillet of the Observatoire National des Effets du Réchauffement Climatique (ONERC) in Paris, from now until 2020 many millions of Africans will be exposed to an increase of water stress because of climate change.[67] In Australia, it is the future of the water reserves of the country that is at stake and people are worried because, with 0.32 per cent of the world's population, the country produces 1.43 per cent of carbon emissions. In 2007, the United Nations Convention to Combat Desertification broke down when its budget was not approved by the US and Japan.[68] Claude Bied Charreton, President of the French Scientific Committee attached to this body, estimates that water stress already affects 1 billion people and 40 per cent of the land.

Thus it is a collection of natural metabolisms that are at stake. Pursuing immediate advantages in favour of a minority, we have brought about profound imbalances and it is possible that they may never be re-established. The idea of unlimited progress is contradicted by reality. Our ancestral wisdom of respect for nature, the source of life, has been vandalized. Exalting one unique value, trading for profit and accumulation, has eliminated other perspectives, and humanity risks paying dearly for it.

On the Economy

From the economic viewpoint, agriculture is particularly sensitive to temperature. In the northern hemisphere, new crops can be cultivated and certain plants have already been introduced. But in the tropical regions the situation is very serious. In India, rice has only 1 °C of tolerance to an increase of temperature; more than this and it has a lower yield (up to 40 per cent) and the same goes for the tropical and subtropical regions as a whole.

Africa is especially vulnerable.[69] Professor Rajandra Pachauri, President of the IPCC, affirmed in the Belgian newspaper *Le Soir*[70] that climate change was already endangering the supply of water of 75–200 million people in the 2020s, as the reduction of cultivable land, of the fertile seasons and of yields will increase the risks of famine. In certain regions, said the professor, the yields of rainfed agriculture could drop by 50 per cent between now and 2020. The UNDP estimates that between now and 2060, economic losses for Africa could rise to 36 billion dollars.[71] As for the climate effects in the Himalayas that we talked about earlier and which are at the origin of decreasing water resources, they could affect a billion people directly in the coming two or three decades.

In China, the economic impact of climate deterioration has been considerable. According to Pan Yue, Chinese Vice-minister for the Environment, the environmental damages are costing the country 8–13 per cent of its GDP each year, thus cancelling out the rate of growth. 'China', he declared, 'has lost almost all that it had gained since the end of the 1970s due to pollution.'[72] Conscious of the problem, the Chinese government decided to undertake, as from February 2008, a national survey on the sources of pollution (which should have been done a long time ago). In fact, electricity production, above all produced from coal, is expected to increase by 30 per cent between now and 2015, and in 2030 between 8 and 10 billion tonnes of CO_2 could be discharged into the atmosphere. (In 2009, the discharge was 6.6 billion tonnes.) It is for this reason that, in 2009, China has been doubling its production of wind energy and expects to overtake the US in this field soon.[73]

China is chronically short of water. With 20 per cent of the world's population, it has only 7 per cent of world water reserves. Cereal production forms the base of the national diet but one tonne requires 1,000 cubic metres of water (also equal to 1 tonne).[74] Julio Arias, author of the *Le Soir* article, adds that the industrial pollution today in China is affecting 90 per cent of the underground water and 700 million Chinese are drinking it.

In many regions the lack of water could also affect the functioning of nuclear plants, which use large quantities for cooling operations. Nicholas Stern stated, in a World Bank report of 2006 that, if appropriate measures were not taken, global warming could bring about the worst economic recession in history. It would cost some 5,500 billion euros – more than the cost of both world wars together or the great crisis of 1929.[75] The economy destined to create the basis of the material, cultural and spiritual life of the human species is being greatly upset by all these phenomena which, let us not forget, are the result of a very specific model of development.

On Society

The social aspect of climate change should also be emphasized. Clearly, it is the poorest populations and regions who will suffer the worst effects, thus challenging the UN Millennium Development Goals (MDGs) (to reduce extreme poverty by half by 2015). Again according to Nicholas Stern, there are likely to be more than 200 million displaced people in the coming decade: those known as climate refugees. Already in Russia, the populations of central Asia are moving westward and the representative of the World Wildlife Fund (WWF) in Russia reports that this region is on the verge of catastrophe. Several Pacific islands are losing inhabitants as they are moving to New Zealand. Eudald Carbonell is even more pessimistic and talks about a social crisis without precedent, saying that during the course of the twenty-first century half of the world's population will have disappeared.[76] Starvation and disease will be the fate of millions of people. First, these will be in the most vulnerable regions inhabited by, according to the IPCC, 'the people least capable to adapt to the climate change because of the increase in malnutrition'. There will probably also be a geographical redistribution of certain vector-borne diseases like malaria, dengue fever and Lyme disease. This is one of the reasons why the pharmaceutical industry has begun to invest more seriously in research for vaccines against these diseases. They will be affecting people with more purchasing power and it is also the reason why the World Health Organization (WHO) is warning the economically better-off regions in the northern hemisphere against these tropical diseases. It will be remembered that during the heatwaves of 2003 there were 70,000 deaths in Europe.

Just as the social consequences of climate change affect all the poor regions, paradoxically those who have less access to harmful technologies applied in the industrial countries and who remain on the margins of the economic growth that serves as a parameter for the capitalist model, are the victims, although they have not been involved as actors in the process of climate destruction. Thus, in 2007, Asia was the main target of natural disasters. This was reported in the balance sheet of the International Strategy for Disaster Reduction, organized by the United Nations: 167 million people suffered the effects of flooding rivers in this region. In the world as a whole there were 16,500 deaths in this kind of disaster and the cost in damages has been calculated at 43 billion euros.

According to John Vidal, 182 million people in Africa could be suffering from diseases linked to global warming.[77] Although they do not attribute all these figures to climate change alone, the competent agencies do, however, state that it has an undeniable influence. Thus Margaret Chan, WHO Director-General, in a report published for World Health Day in 2008, affirmed that air pollution was responsible for 80,000 deaths each year and that each rise of 1 °C in temperature would mean 20,000 extra deaths. The figures issued by the Global Humanitarian Forum in Geneva on 26 June 2009 are even higher. They calculate that, in 2008, all the natural catastrophes have caused 315,000 deaths and have affected 325 million people. In 2030 the statistics could be respectively half a million deaths and 20 per cent of the population affected.

In 2007 Africa was devastated by torrential rains and floods that caused at least 150 deaths. Seventeen countries were affected and a total of 1 million people, according to the United Nations Office for the Coordination of Humanitarian Affairs. In Uganda, 150,000 people lost their homes between August and September and 400,000 lost their means of subsistence. In Sudan there were the worst floods in human memory. In Togo some 66,000 people were displaced, and in Nigeria, 50,000. All this was due, according to the same agency, to climate change and the destruction of forests.[78]

The populations of the Pacific islands are particularly affected. They are already suffering from food and water shortages, more frequent flooding and storms, while malaria is on the increase. According to a recent Oxfam report, more than 75 million people will have to relocate by 2050 because of the effects of climate change, if nothing is done to ease it. Many will not be able to relocate in their own islands and will become international refugees.

But what about the 'emerging' countries? Let us look at China again. This is a society that has opted for accelerated growth based on the polluting technologies which, in the short and long term, have frightening effects. We should of course look at it in a global context. After a period of economic and political dependency, which lasted for centuries, the emancipation of the nations of the South seemed like reparation for injustice. The North had an economic development model based on capital into which it integrated the extraction of natural resources and the exploitation of work in the South. It was hardly concerned about the natural and social consequences of its growth practices and is therefore chiefly responsible for the current ecological situation. In the eyes of the South, particularly of the G77 (in other words, the 'developing' countries), the North cannot give lessons to anyone.

Unfortunately, the ecological price to pay does not take into account historical justice and a fair return to an equilibrium. We are all in the same boat and the folly of some cannot justify the heedlessness of the others, even if there are varying degrees of responsibility. As Susan George says, we are all on the *Titanic*, even if some are in First Class. We know the cost of an accelerated predatory development in a world saturated with micro-particles and greenhouse gases, and China, more than the other emerging countries, is in the process of paying for it. The pollution of its large cities has become legendary. The Olympic Games in Beijing revealed it only too well.

Let us take Hong Kong. Each year several hundred people die because of polluting emissions. The index of air pollution, measured by the number of fine particles in suspension, has been fixed by the WHO as a maximum of 20 micrograms per cubic metre ($\mu g/m^3$) of particles with a diameter less than 10 microns, i.e. one-hundredth of a millimetre (PM10). The WHO even recommends not exceeding 10 $\mu g/m^3$. This index has been changed in Hong Kong so that the tolerance levels are far higher. The acceptable level established by the city is 180 $\mu g/m^3$ and the official threshold has been raised to 380 $\mu g/m^3$, even arriving at 800 $\mu g/m^3$, as long as it does not last for more than one hour. In Europe, since 1 January 2005, it is not allowed to exceed 50 $\mu g/m^3$ for more than 35 days a year and an annual limit is fixed at 40 $\mu g/m^3$. Florence de Changy, who quotes the figures for the Chinatown of Hong Kong, has no problem in writing that these units have been adapted 'so as not to upset the interests of the local industrial interests'.[79] Once again capitalist logic prevails. As long as the effects only concern 'externalities' and do not directly affect the process of the accumulation of capital, they are not taken into account.

However, in June 2003, a group of university staff in the city was worried about the problem and published a study concluding that an improvement in the pollution index would relieve the hospitals of 36,000 patient days a year, thus economizing 1.9 billion euros of indirect costs. But as these are met by the public authorities or individuals they do not enter into company accounts. Moreover, as many of the hospitals are private, there would hardly be an advantage in reducing the number of their clients. For China as a whole, the World Bank calculates that atmospheric pollution creates hundreds of thousands of victims each year. It must also be responsible for the fact that each year a million babies are born with deformities. At the world level, it is an economic development model which entails sacrifices for objectives that are increasingly removed from human welfare – China evidently is not the only one.

In 2008, in India, nearly half the population still had no access to electricity and it has been estimated that the demand for energy will quadruple in that country over the coming 25 years (more or less by 2030). Even if this has a negative impact on the climate, we should keep a sense of proportion and remember that the average American citizen emits 16 times more CO_2 than an Indian.[80] According to Sunita Narain of the Centre for Science and Environment of New Delhi, the revolts that have taken place in recent years in 200 of the poorest districts of the country are due to the destruction of forests, pollution, falling levels of water supplies, and the extraction of mining resources. The same author believes that climate changes will have dramatic social effects, caused by increasingly scarce water, diminished food security and reduced land in the coastal areas.[81] Forced migration will increase, with all the problems that this entails. Briefly, it is the collective well-being of humanity that is at stake and we must be aware of this.

On Politics

The Oxford Research Group has published a report entitled *Global Responses to Global Threats* that declares that climate change – displacement of populations, lack of food, social disorders – will in the long run be much more important for security than terrorism. It is an opinion that is shared by the Pentagon itself. Lester Brown talks of all the non-viable states that are incapable of administering their problems concerning water, forests, soil erosion, migrations – in sum, the effects of climate change. The effect of climate migration also risks becoming very serious: the criminalization of migrations in Europe, the construction of walls between the US and Mexico and between India and Bangladesh.

We have already described the effects of global warming on the Arctic, making it possible to exploit its rich natural resources (oil, metals) and the opening up of a new sea lane. Canada claims sovereignty over it, while the US and the EU affirm that it constitutes international waters. Canada is constructing a military base in the region. The US proposes signing the Law of the Sea, put forward by the UN in 1992, which represents a veritable turn-around in its own position. In fact, starting in 2009, it is envisaged that countries can extend their territorial waters to more than 200 miles if they can prove that their continental shelf extends that far.

As for military activities that influence the climate, these should not be underestimated. Not only does an army, because of its use of energy, impact considerably on the ecology (in France its production of nuclear waste is responsible for 10.1 per cent) but the ecological cost of wars is immense – for example Iraq. But there are many other consequences of military activities. In South California, the use of powerful sonars by the US navy, in preparation for an underwater war, is seriously affecting the brains and ears of whales and other marine fauna.[82] In Puerto Rico, after 60 years of US army exercises at Vizques, it will take ten years just to remove the unexploded shells from the territory.

The military has understood that these days the protection of the environment can give it a positive image among the public. The Spanish Ministry of Defence had full-page publicity in the country's newspapers on 13 December 2007 in which it stated that 'The natural milieu is also an issue that concerns defence. In the Spanish state, 33 [military] zones cover 150,000 protected hectares.' The Ministry added that there were many ways of defending the territory, that being concerned about the flora and fauna was one such way and that by applying an environmental management system to all its units and installations and opting for renewable energies, the army is fulfilling its duties so that everyone can benefit from the environment. The publicity ended by saying: 'For your future and for that of everyone!' As for the US air force, it has had numerous experiments in using coal-based fuel for some 40 planes (for example the B52, the C17, etc.).[83]

What is even more worrying are the possibilities of modifying climate change for military purposes: what Juan Gelman terms 'the climate war'. It is becoming possible to discharge high-frequency waves into the econosphere in order to provoke violent rains and flooding, upsetting communication systems, as the US army is at present experimenting in Alaska.

All these effects – ecological, economic, social and political – make impressive reading. So it is understandable that Enrique Leff, at the National University of Mexico, concludes that humanity itself is facing a crisis. He believes it is the development model that is at issue and that as long as the question is not posed in these terms, only the effects will be dealt with, not the causes.[84] For its part the

UNDP talks of a tragedy in preparation,[85] and Lester Brown of a 'collapse of civilization'. In this situation, the decision of the G8 in Aquila in June 2009 to reduce CO_2 emissions by 80 per cent between now and 2050, without fixing interim targets (and with the disagreement of Russia), seems lamentably inadequate.

The Response of the United Nations

The UN could not remain indifferent to the problem, even if the orientation of its collective actions obviously depends on the consciences of its members. Back in 1986, the Brundtland Commission was already referring to it, as was the Earth Summit at Rio de Janeiro in 1992. The two most well-known UN climate meetings were the ones at Kyoto in 1997 and Bali in 2007. The Protocol signed at Kyoto by most of the industrially developed states in the world, but not ratified by the US, Japan and Australia (this last rectifying its position in 2007), aims at reducing by 6.5 per cent (compared with 1990) the emissions of the six main elements that are altering the climate, between 2008 and 2012. The developing, including the emerging, countries (China, India, Brazil) were not affected by these measures. The 13th Conference on the United Nations Framework Convention on Climate Change (UNFCCC) took place in Bali in December in order to prepare the follow-up to Kyoto. It was followed in December 2008 by a meeting at Poznan, prior to the holding of the conference at Copenhagen in December 2009, which has to take decisions that will bind member states.

The first conference, at Kyoto, laid down the norms for the reduction of emissions, coupling them with rights to compensation, which led to the project of carbon exchange about which we shall talk later. The second, at Bali, was prepared by several meetings of the IPCC, the body created jointly by the WMO and the UNEP, which is holder, together with Al Gore, of the 2007 Nobel Peace Prize. The Bali meeting took place in a distinctly more dramatic atmosphere: the US, increasingly isolated, was obliged to make certain concessions while forcing others on the rest of the assembly – although of less importance. As for Australia, with a labour government in power, it had joined the majority camp. However, all this did not prevent the industrial lobby from being extremely active in trying to influence decisions in a way that favoured market forces. At the same time, in compensation, there was a drawing together of the organizations and movements concerned with the environment and those that defended social justice and human rights. This was welcomed as a great step forward by Walden Bello of the Focus on the Global South of Bangkok. The following stages, the 14th meeting of the UNFCCC in Poland in 2008, and the 15th meeting in Denmark in 2009, should lead to new concrete orientations for the period after 2012. The most important concern is to reduce CO_2 emissions from 20 per cent to 40 per cent compared with 1990 levels.

The emerging countries were questioned at Bali. Together, South Africa, Brazil and China all agreed that they were genuinely concerned. China announced its willingness to reduce its energy consumption by 20 per cent by 2012 and it announced its decision to put a tax on the exportation of wheat, maize and soya. Several countries from the South, Brazil, Indonesia, Ecuador and Mexico, asked for financial support so that they could conserve their forests. China pleaded for a free transfer of technologies that reduced greenhouse emissions. In the same spirit an Adaptation Fund was created, to be administered by the Global Environment Facility and endowed with 500 million dollars by 2012, with the money coming from 2 per cent of the contributions for projects reducing CO_2 in the framework of the Kyoto Protocol in order to finance green energy projects in the Third World. However, the sum is paltry compared with what is necessary, calculated to be 50 billion dollars by Oxfam and 86 billion by the UNDP. The World Bank, according to François Bourguignon, Director of the Ecole d'Economie de Paris, would become the Environment Bank[86] for the occasion and would organize the modalities of the funding. There is little doubt that, if this happened, the initiatives would remain 'market friendly'.

Before the Bali conference, 125 NGOs published an appeal demanding, among other things, that not only economic criteria be taken into account, as there were other criteria of well-being. They wanted the GDP measurements to be replaced by another kind of measurement that would take the impact on the climate into consideration. And they also demanded the creation of financial resources to protect the forests of the poor countries, the setting up of an international fund for clean energy and the adoption of a Convention on the Right to Water. The UNEP, in its report of March 2009, entitled *A New World Green Pact*, shows how a 'Green New Deal' could relaunch the world economy, create jobs and protect the most vulnerable groups.

There are other agencies of the United Nations involved in these fields, like the WMO, the Sustainable Development Commission, the Convention on Biological Diversity, the Global Environment Facility, which groups 160 countries, the UN Convention to Combat

Desertification and UN Habitat, not forgetting the UNDP, the FAO and the United Nations Conference on Trade and Development (UNCTAD). To these should also be added a certain number of regional bodies such as the European Environment Agency (EEA).

The UN framework provides a certain space for the autonomy of those who, concerned by the situation, are trying to raise awareness of the importance of the problem and of the need for radical solutions. But the balance of power comes into play, not only in the political field, but also with regard to the concepts of development and economic philosophy – in a word, a project for society. Hence Tony Blair, who declares that the challenge is immense and the time limited, warns of the risk that the Copenhagen summit in December 2009 will reach an agreement on the lowest common denominator, as each country will concede as little as possible.[87] In fact, neoliberalism has not yet had its last word and all the main orientations of the world economic system, even shaken as it is by serious crises, are still subordinated to their original logic.

This leads us to wonder about the neoliberal discourse on the climate and the effects of global warming: how it is interpreting the situation, the solutions it is proposing and the logic guiding their formulation. In fact, the climate phenomenon has now acquired such high visibility that it has become impossible to ignore its dimensions, even if certain voices are still minimizing it or denying the pertinence of the available data and their interpretation. By examining all this we shall understand better the function of agrofuels in the reproduction of the economic model.

3
The Neoliberal Discourse on Climate Change

In spite of increasingly universal acceptance of human responsibility for climate change, there are those, like former President George W. Bush who did what he could to conceal or minimize it. But there are various ways of approaching the subject and it is worth studying the different approaches of the neoliberal discourse.

Scepticism

First there are the sceptics that question both the value of the data and their interpretation. The complexity of the models and the intrinsic unpredictability of the climate arouse suspicions in the likes of writers such as Bjorn Lomborg in Denmark and Claude Allègre in France. Indeed, the state and results of meteorology could justify such distrust. Nature can be chaotic, observed Pascal Engel, a philosopher from the University of Geneva in an article in *Sciences et Avenir*[1] *apropos* 'climate scepticism'. But is this caused by meteorological factors or by disturbances created by human activity? These are certainly measurable, but the same cannot be said with such certitude about all the effects. This is a key question. In a book on the subject, Professor F. Ruddiman of the University of Princeton is not overly worried. He tends to the first interpretation. The process has been going on for a long time, he writes. It started some 8,000 years ago, with the deforestation caused by the development of agriculture.[2] The first reactions of the liberal discourse rested on affirmations of this kind and relativized the importance of the phenomenon.

At the beginning of the 1990s a powerful lobby was organized to prevent an international treaty on the reduction of greenhouse gas getting under way. Not long before the Earth Summit was held in Rio de Janeiro in 1992, the George C. Marshall Institute, a US conservative think tank, published a report saying that climate change was probably the result of excessive solar activity.

When the NASA climatologist James Hansen stated, in June 1998, before the US Senate that he was 99 per cent convinced that greenhouse gas was in the process of changing the climate, the industries involved became still more worried and set up several pressure groups, such as the Global Climate Coalition (GCC) and the Committee for Environmental Information. The latter stated that global warming was a theoretical hypothesis, not a reality. As can be seen, it was a strategy similar to that developed about cigarettes and the effects of tobacco. During the 2002 Earth Summit (Rio plus 10) in Johannesburg, there was a lobby that was particularly active trying to minimize the situation, the World Council for Sustainable Development, constituted by some 190 oil, chemical and timber companies.

One of the most frequent arguments was that the scientific community was divided. The Goddard Institute for Space Studies (GISS) stated at the end of 1990 that the hottest year in the US had been 1934 and not 1998, as had been said and that the six hottest years in the meteorological history had occurred in the US between 1930 and 1940. However, the same institute had also to acknowledge that the temperature had increased by 0.21°C since 1920. But it is not only in the US that such interpretations were put forward. In the framework of the Institut de Physique du Globe in Paris, Claude Allègre, whom we have already mentioned, Vincent Courtillot and Jean-Louis Le Moël believed that there were other causes rather than human activity to explain global warming, such as variations in solar activity, the intensity of cosmic rays and the oscillation of the earth in its orbit.[3] The argument was similar with all of them: emphasizing natural causes, so as to help minimize human causes.

An editorial in *The Economist* of 2 June 2007 clearly stated that up until recently the business world tended to be suspicious of statements about climate change. In fact, such a notion implied that the industry had endangered the planet and that it should therefore pay the consequences. As that was out of the question, it was better to deny the facts. Today, added the journal, everything has changed. Everyone is rushing to parade their 'green' performances. Even better, 'cleaner energy means new technologies and greater possibilities of making money'. Etienne Davignon, the Belgian financier, confirms it: 'Environment and economy are already integrated. The environment is no longer a separate chapter.'[4] This conviction was shared by Jeff Immelt, the CEO of General Electric who, in 2005, declared 'From now on business and ecology will have a win–win relationship.'[5]

It was not only a verbal offensive. It was also necessary to fend off the conclusions of the research carried out by scientists. Thus Exxon Mobil spent more than 10 million dollars to finance centres in the US who had to prove that global warming was just an illusion or a recurring phenomenon – at all events, nothing to worry about. Among the 20 or so institutions concerned was the American Enterprise Institute which, in 2004, published a study on global warming entitled *Don't Worry, Be Happy*.[6]

The resistance of the economic world also had repercussions in the political arena. Thus former President George W. Bush, in his first mandate, nominated former coal and oil lobbyists to key posts in the formulation of climate policy. Sharon Begley describes in the *Wall Street Journal* the reactions of the US governmental authorities to the alarmist report of the IPCC in 2007: 'It was not enough to offer 10,000 dollars to scientists ready to criticize the text, they put forward a new theme: even if the earth was warming up, there's no reason to be alarmed.'[7] But the White House did not stop there. It arranged modifications in the report of Dr Jules Gerberding that was presented to the Senate and concerned the impact of global warming on health. Passages about the diseases that risked developing in the future were suppressed.[8]

The most striking case was that of James Hansen, Director of the Goddard Institute for Space Studies. When, in 1989, he appeared before the Senate Commission presided over by Al Gore he affirmed that paragraphs had been added to his written statement that contradicted his own conclusions. This was revealed in Mark Bowen's book *Censuring Science*, published by Dutton Books in the United States. But the harassment continued for years afterwards. The same researcher wrote, in 2005, 'Censorship has become very intense, prohibiting me, for example, from writing in the media', and later, 'I was not even able to put on-line the temperature statistics as I used to do every month over the last dozen years.' As Bowen explained, the White House was determined to control everything that came out of his laboratory.

Then there was also the case of the researchers at the National Oceanic and Atmospheric Administration (NOAA).[9] And, as late as June 2008, the Republican minority in the US Senate was still creating problems for the first important legislation for fighting global warming.[10] However, contrary to the Bush administration, that of Barack Obama shows signs of an opening in favour of ratifying the Kyoto Protocol. On 26 June 2009 the President presented his project to the Senate. He envisages a reduction of

17 per cent in CO_2 emissions between now and 2020. After the negative vote of 212 parliamentarians, Paul Krugman did not mince his words when he spoke, in an editorial in the *New York Times*, of the 'immorality of denying global warming'.[11]

In Europe, too, the industrialists were organizing to reduce the objectives of the EC while the pact on the climate was being prepared at the beginning of 2008. As Philippe Régnier wrote, 'Never has the lobbying of big industry on the European institutions been so powerful.'[12]

Delegitimizing the Scientific Evidence

As well as scepticism and political manoeuvres there was also delegitimization of the scientific approach, and here there was no lack of epithets. Let us start with a long quote from Eric Le Boucher in *Le Monde* of 9 April 2007:

> To raise awareness, the IPCC politicians and the professional ecologists want to make people afraid and to do this they simplify, if they do not amplify. Why do they need to pretend they are economists and predict recessions in the United States? Why state with such certainty that 2 °C more would be bad for trade, when we can see in the United States itself there is a move of activities towards the south and the sun? To say that 'it will be the poor who will suffer more from global warming' touches our emotions. But it is a truism. The poor suffer more also from a cooling of this planet simply because the world is made like that – it is always the poor who suffer most, whatever happens. It has been said here already and we shall say it again: this strategy of the 'ecolos' is disastrous for the ecological cause itself. The scholars of the climate would do better to refine their studies and not to be militants. Politics has never sat happily with truth.

It would be quite difficult to be more contemptuous.

The American magazine *Newsweek* dedicated its number of 16–23 April 2007 to the climate theme. To oppose ultra-pessimism becomes a heresy, write Jonathan Adams and Kenzie Burchell, and they quote Thomas Moore when he said: 'In more religious times, we used to say that we were sinning against God. Now we say that we are sinning against nature. Moderate voices are being drowned out.'[13] As the journalists say, we are in an 'eco-puritan' era. Regarding the report of Nicholas Stern for the British government, he is described as 'semi-apocalyptic' by Emily Vencat in the same issue,[14] while Mac Margolis has no hesitation in speaking about the 'Climate Cassandras'.[15] Richard S. Lindzen, meteorologist professor at MIT, goes even further:

Recently many people have said that the earth is facing a crisis requiring urgent action. This statement has nothing to do with science. There is no compelling evidence that the warming trend that we've seen will amount to anything close to catastrophe ... A warmer climate could prove to be more beneficial than the one we have now.[16]

Adams and Burchell echo him and add: 'Rising temperatures might even be good news for some ... Previous warm spells, notably in the Middle Ages, are associated with prosperity and the advance of civilization.' They give as examples the opening of the new sea-lane in the Arctic and that in Canada, Russia and Scandinavia the benefits of greater heat would mean more abundant crops, fewer winter deaths, reduced heating costs and a possible boom in tourism and housing.[17] Richard S. Lindzen had no qualms about responding to Al Gore that the heating of the planet was a huge swindle, adding that the increase in temperature that India experienced during the second half of the twentieth century enabled it to have a large increase in agricultural production. Besides, he says, 'exposure to cold is generally found to be more dangerous and less comfortable'.[18] In brief, everything was done to minimize the extent of the problem.

SECOND STAGE: PROMOTING MARKET-FRIENDLY SOLUTIONS

As the extent of the climate crisis is ever more evident it becomes increasingly difficult to continue to proclaim that it does not exist. So we enter a second phase, in which market-oriented solutions are put forward.

Optimism

There was an extremely fast turnaround. Optimism began to prevail in the neoliberal discourse, based on confidence in the progress and capacities of human knowledge. Thus Peter Levene, director of Lloyd's of London, which has been hard hit by recent natural catastrophes, declared: 'The human race invented air conditioning and central heating. We will adapt to these new conditions.'[19] Mac Margolis has affirmed that 'civilization's future now rests largely on [farmers'] ability to decipher, harness and adapt to the climate of tomorrow. Fortunately farmers have been doing just that for millennia.'[20] And the neoliberal discourse cites the archaeologist of University College of London who said 'For every society that collapses, there is another that consolidates.'[21] Philippe Manière, Director of the Institut Montaigne in Paris, considers that the genie of capitalism is its capacity to adapt:

> In the long term I am very optimistic. The history of humanity is made up of very serious problems that have been solved. In five, ten or fifteen years' time there will be unprecedented forms of growth with different sources of energy, of individual and collective means of transport functioning according to an alternative model.

Jeffrey Sachs writes:

> Numerous ideologues of the free market pour ridicule on the idea that the constraints imposed by natural resources could cause a slowing down of world growth. They say that the fear of scarce resources, particularly food and energy, has been with us for 200 years and we have not succumbed. On the contrary, wealth continues to increase faster than population.[22] ... That is the genius of capitalism: it continually adapts to the new facts based on experience.[23]

The development of renewable energies in particular can only be compared with the industrial revolution, say the European experts.[24] 'Natural capital has been replaced by industrial capital.' The turn to new energy, they say, has already created 100,000 jobs and generated a business of 20 billion euros (especially in the field of agrofuels). In sum, optimism is the order of the day. Solutions will be found and, what is more, in the framework of the present economic system, which has always promoted innovation. Besides, 'the cost of inaction would be greater than action: it would therefore be more expensive to repair the damages of global heating than to invest in order to avoid it'.[25]

It is true that industry is adapting rapidly to the new situation. Efforts to reduce energy consumption turn out to be worthwhile. Energy efficiency has improved by 1.6 per cent a year since 1990, reducing the CO_2 emissions in the atmosphere, if only for 2006, by 10 Gt, according to a report of the World Energy Council. For comparable service we consume about 40 per cent less energy than

15 years ago. In Germany, CO_2 emissions have been reduced by 20.4 per cent between 1990 and 2007, half of which has been caused by the collapse of industries in East Germany. In Belgium, there was a reduction of greenhouse gases of 6 per cent in 2006 compared with 1990. These results are due to the efforts of both households and industries: improvement in the rate of energy yields, more efficient technologies as a result of higher prices of energy and incentives by public authorities.[26] Nevertheless, in Europe, 2007 saw a rise of 1.1 per cent of CO_2 due, according to the EC, to the application of the rights to pollute (see below).

In the US, metallic sponges (ZIFs) have been developed that are capable of absorbing 83 times their own volume. They can serve as a filter in the chimneys of coal plants, to then be cleaned of the accumulated CO_2, which can finally be buried. However, the high cost of this solution, which joins cobalt and zinc with organic molecules, prevents it from being generally applied for the moment. All this should remind us, as Anthony Giddens emphasizes, that the reduction of CO_2 emissions is not a synonym for a slowing down of the economy and that, on the contrary, it constitutes an opportunity for the creation of new green jobs in the various sectors of the environment.[27] This is confirmed by UNEP.

We should remember that the capitalist economy logic has a double principle: first the cost-benefit ratio (calculated without externalities) must be in favour of the benefit, and then competition demands that the costs of production are not increased, at risk of disappearing from the market. These rules also apply to the measures to be taken to stop the destruction of the climate: short- and medium-term benefits should be assured and policies avoided that could profit rivals less careful about costs. Once again, it is when the rate of profit on capital risks being halted or paralysed that economic actors adopt what are held to be 'rational' measures, in function of the system's logic. In the short term, therefore, only general regulation imposed by public authorities can resolve the problem. But this also risks being conditioned by the danger of competition from countries that possess the 'comparative advantage' of less restrictive ecological legislation, like the threat of industrial delocalization when this becomes more profitable for capital. Countries are subject to the law of the market and governments, even social democratic ones, feel obliged to defend their companies internally and externally, in contradiction with climate protection.

However, ecological concern also becomes a commercial argument. We see it with the automobile industry which, in its publicity, stresses the less polluting character of its latest models while the EC wants to reduce CO_2 emissions to 120 grams per kilometre by 2012. Renault's Logan Eco2 car was presented at the Shanghai salon in November 2007 as having an atmospheric discharge of only 136 grams of CO_2 per kilometre, far less than previous quantities. Volkswagen, which has acquired a lot of territory in Brazil and made an agreement with ADM (Archer Daniels Midland) to produce agrodiesel, launched its Bluemotion by presenting it as the least polluting car on the market (102 grams of CO_2 per kilometre). The company in Spain has pledged to plant 17 trees for each new car sold which, it says, makes it possible to absorb the CO_2 discharged during the first 50,000 kilometres the car is driven. The PSA group in France presents both the Blue Lion (Peugeot) and the Airdream (Citroën) that emit less than 130 grams of CO_2 per kilometre with fossil energy. Ford has not been left behind, with its Econetic and the ecoboost hybrid engine that equips the Ford Explorer America. Mercedes announces a 'bionic car' joining high technology and low CO_2 emissions. Opel, of the General Motors group, stresses the 'flextreme' technology in which propulsion is created by electricity, the petrol or diesel engine only serving to recharge the batteries.[28] The Toyota Lexus is a hybrid car, using petrol and electricity. The company has made an agreement for the distribution of ethanol with BP, Peugeot, Citroën and Saab. Better still, Mitsubishi has launched the electric mini-car Miex, which is advertised as having zero emissions, while Daimler has bought part of the Californian company Tesla to make electric vehicles.

Advertisements go further still. Thus Peugeot encourages people to buy the new models with this statement: 'Twenty per cent of the oldest cars are responsible for 60 per cent of polluting emissions from cars. Replace them!' It has to be said that the European requirements are ambitious: by 2012 we must not emit more than 120 grams of CO_2 per kilometre, whereas in 2007 we were at 160 grams, according to the European Association of Car Makers (ACEA). This body thinks that it will be difficult to achieve the aim of reducing emissions to 140 grams for 2008.

Evidently there is no objection to producing cars if they are less polluting. But it should be remembered that the increase in automobile production at the world level cancels out the reduction obtained and that the ecological concern has only been demonstrated recently, while car pollution has been denounced by scientists and ecologists for decades now. We had to wait for international measures to be announced before it became a publicity argument.

As Carlos Migueles has written: 'With such publicity, the brands very much hope they will recover the costs of what they have had to spend.'[29]

There are other initiatives, highly publicized, to show the concerns and efficiency of the global actors. Monsanto has announced that by about 2015 it will have cotton seeds that are resistant to drought, in view of the coming climate change. General Electric reinforces the energy efficiency of its products.[30] Ethanol will be produced not only by the US and Brazil, but also by Canada and Russia. Insurance companies are inventing new products: Katrina – the hurricane that devastated New Orleans – has laid the ground for a veritable 'new market' in this field.[31] Yale University and the Ceres think tank organized a seminar with the directors of 1,000 companies, designated by *Forbes* magazine as the most efficient in the world, on the risks and opportunities of climate change.[32]

In Africa, intelligent policies are being implemented to remedy climate calamities, writes Pedro Sanchez of Columbia University in New York. Thanks to private foundations, he explains, particularly those of Bill and Melinda Gates and the Rockefeller Foundation, a burgeoning 'green revolution' is in process in Africa: 'enriching barren soil, training farmers and providing them with hardy hybrid seeds and working with the private sector to help farmers enter the market place'.[33] In brief, the genius of the capitalist system is to transform catastrophes into opportunities and tragedies into profits when the private sector is given the leadership in humanitarian intervention.

Towards the end of 2007, the French paper industry published a full page advertisement in the country's daily newspapers (25 October 2007):

Yes. Paper is contributing to the struggle against greenhouse gas by using wood from forest clearings, replanted plantations and off-cuts of wood. The making of paper promotes the growth of trees and thus enables the absorption of carbonic gas. Once paper is made, then recycled after use, it continues to capture the carbonic gas contained in its fibres.

The deceptive impression of this statement comes less from what it says than what it does not say. In fact, what is the contribution of paper produced from eucalyptus plantations to the ecological and social disasters in the tropical regions? With what kind of energy is paper produced? How much CO_2 does it discharge into the atmosphere when it is being transported? One cannot isolate one factor or else a false impression is given. But ecological fibre is now booming and therefore the market takes it over. Even McDonald's claims in its publicity that it is making children aware of eco-citizen actions.

This brief presentation of some opinions, to be sure eclectic, but significant, aims at introducing the reader to the great diversity of arguments of a neoliberalism that is faithful to its original logic. We shall conclude with one last quotation: it is from John Llewellyn, senior economic adviser at Lehman Brothers. He sums up the basic philosophy very well: 'Global warming is likely to prove one of those tectonic forces – like globalization or the ageing of a population – that gradually but powerfully changes the economic landscape.'[34] Such a perspective sees it as a challenge that creates opportunities to be seized by those who are capable. They will find the appropriate technologies and thus be able to pursue their capital accumulation. This was echoed in an advertisement in the American newspaper USA Today of 24 April 2007: 'How global warming can make you wealthy'. Faced with this avalanche of optimism, the climatologist Jean Pascal van Ypersele, Vice-president of the IPCC, had to conclude: 'the discourse of the economists who present a radiant world in which technologies that are economic in CO_2 will cost less, is dangerous'.

Respecting the Laws of the Market

Are we saying then that for the actors of the neoliberal system, nothing should change? Obviously not. However, for all the neoliberal writers and most of the social democrats there is one fundamental condition for any adaptations: that they respect the laws of the market. So says Nicholas Stern of the London School of Economics (LSE) and author of the report on climate change commissioned by Tony Blair, when he stated that there had to be agreements that associate the gains of Kyoto with the mechanisms of the market to implement the transition of development models.[35] Thus the Kyoto agreements envisaged exchanges between polluting industries and developing countries and quotas of CO_2 thus enabling greenhouse gases to enter into the field of exchange values. London has therefore developed a stock exchange in CO_2 which has become not only a booming commercial activity, but is the guardian of the prices for the capture of CO_2 by regions that are rich in biodiversity.[36]

The system envisages that each country can dispose of a quota of 'assigned amount units' (AAUs) of gas emissions authorized for five years between 2007 and 2012. This date was fixed by the Kyoto agreements in order to arrive at a reduction of 5.2 per cent of greenhouse gas by comparison with 1990. Some countries with too many emissions will be above their quota, and others will be below theirs, and they can make exchanges. Three mechanisms have been envisaged. First, trade between countries or what is called 'the sale of hot air'. A state that is emitting too much greenhouse gas can buy quotas from another state that has not reached its established limits. However, it cannot be for more than 10 per cent of its allocated quota and this is monitored by the secretariat of the UN Climate Change Convention.

A second mechanism is the Joint Implementation Project which enables industrialized countries to implement emission reductions in another industrially developed country at a lesser cost than at home. For example, the countries in Western Europe can join with countries in the former socialist countries of Eastern Europe. 'Emission Reduction Units' (ERUs) can then be put to their credit. The third is the Clean Development Mechanism which concerns the action of these countries in the developing regions. In 2007 this represented a market of nearly 10 billion dollars. Initiatives to replant a forest, for example, can make a contribution to the national allocation quota. Everything is to be strictly monitored internationally, which does not prevent a great deal of manipulation. In December 2007 there were more than 2,000 projects and the aim was to reduce CO_2 emissions by 2 billion tonnes between now and 2012. In spite of everything, after that date, it would still be necessary to halve the quantity of emissions in order to keep the increase in global temperature below 2 °C.

One can see why an important emissions trading exchange (also known as 'cap and trade') has developed. In 2006 there was a collapse in the price of CO_2 per tonne, probably because policies were too generous, but on the whole the operation has been remarkably successful. That same year the market handled some 20.5 billion dollars and this sum could reach 80 billion dollars in a few years' time.[37] However, this enables industries to meet the reduction of emissions without lowering their greenhouse gas emissions.[38] As we have already said, this meant, in Europe, an increase in the emissions of CO_2 by industry. Nevertheless, it is expected that the terms will be more severe after 2012. In 2007 the price per tonne was fixed at around 24 euros.

New companies are specializing in this field, such as Low Carbon Accelerator, Powernext Carbon, European Carbon Exchange, etc., and eco-business is in full expansion. According to the German Minister for the Environment, the four largest European energy producers have earned between 6 billion and 8 billion euros.[39] In January 2008 the EC proposed setting up an internal European market for the 12,000 most energy-consuming companies, with a possibility of auctioning as from 2013. The EU, for its part, set up the Greenhouse Gas Emission Allowance Trading Scheme with the same objective.

As for the UN, during the negotiations in Bonn in June 2009, it proposed to adopt national targets in the carbon markets, in the form of reduction initiatives adapted to the different countries. It should, however, be mentioned that in Sweden, since 1991, a carbon tax has been established that has diminished the levels of emission by 9 per cent between that date and 2008, while economic growth has grown by 48 per cent.[40]

It is interesting to note that in certain American circles such a system is criticized, not only because it risks being a source of corruption, dumping and blackmail – which is also a real possibility – but above all because it imposes a constraint on economic activities. According to them, measures respecting market-friendly principles, such as the taxes on CO_2 and the lifting of taxes on renewable energies, would provide more stimulus rather than impose restraints. Hence, in Minas Gerais state in Brazil, eucalyptus plantations destined for charcoal count as 'reforestation'. A study in April 2007 by David Victor and Michael Wara of the University of Stanford in the US also showed that there have been unjustified gains by industries that have destroyed an industrial activity which had high greenhouse gas emissions (HFC-33). The Clean Development Mechanism had enabled them to receive 4.7 billion dollars, while the cost of destruction was probably less than 100 million dollars. Of a similar opinion is Daniel Esty of the University of Yale, in his book *Green to Gold*, which is quoted by Fareed Zakaria.[41] At the political level the same argument was utilized by George W. Bush for refusing to ratify the Kyoto Protocol.

One could argue that everything that contributes to reducing the emission of greenhouse gases is welcome and therefore also this kind of initiative is acceptable. But is it not going too far in ignoring that this type of solution first of all follows the logic of reproducing the economic system? And not as a response to the problem of civilization, which is after all the fundamental issue? For example, the American transnational corporation Arcadia Biosciences plans to finance the development in China of GM crops (particularly of rice) through the compensations envisaged by the carbon trading market. The reasoning is that, with the GM seeds supplied by Arcadia, peasants will use much less fertilizer that emits nitric oxide. As this is 300 times more harmful than CO_2, with the sale of quotas of CO_2 the peasants could finance the purchase of the GM seeds. This experiment is due to be implemented in the province of Ningxia in the north of China.

But the company's plans do not stop there. After rice, there could be wheat, maize, sugar cane, beetroot, oleaginous plants, cotton – even lawns for golf courses. As agriculture contributes more to greenhouse gas than transport (between 14 and 17 per cent, according to different calculations), the company very much hopes that the Chinese experience, programmed between now and 2012, can become generalized and applied to other continents. It has already sold its patent for oleaginous plants to a number of other companies, including Monsanto.[42] The *Wall Street Journal* of 10 October 2007, which also reported this development, added a clarification by the director of the Californian corporation that the problem in China is that the traditional mind of the peasant and the lack of respect for intellectual property rights make it difficult to make money. It is also the reason why Monsanto earlier had already thrown in the sponge. The newspaper added that China is still reticent in using GM seeds other than for cotton.

The group Global Leadership for Climate Action (GLCA) follows the same kind of logic. This initiative is an outcome of the Club of Madrid, which groups some 64 former heads of state or governments 'to reinforce democracy in the world', together with the United Nations Foundation, a public/private initiative of the media mogul Ted Turner to 'take up world challenges'. Members of the Club of Madrid include Gro Brundtland, the former prime minister of Norway, who was also President of the first United Nations Commission for Sustainable Development; Fernando Enrique Cardoso, former President of Brazil; Enrique Iglesias, former President of the InterAmerican Development Bank; Lionel Jospin, former French prime minister; Ricardo Lagos, former President of Chile; George Soros, the US financier; the above-mentioned Ted Turner; James Wolfensohn, former President of the World Bank; and Ernesto Zedillo, former President of Mexico. The group presented a report entitled 'Framework for a Post-2012 Agreement on Climate Change' to the G8 in Berlin on 11 October 2007.

In its introduction the report says that such an agreement must be 'in line with the norms established for economic growth and sustainable development and integrate poverty alleviation in its strategies'. The report continues by emphasizing public/private partnerships: 'the private sector is best equipped to make incremental improvements in the deployment and diffusion phases ... However governments need to offer clear and predictable frameworks to support deployment in their countries.'

The report is signed jointly by Ricardo Lagos and Timothy E. Worth. Former democrat candidate for the US presidential elections, John Kerry, says much the same thing. Solutions must 'give guarantees to the market', he declared at Bali during the UN Conference, and he singled out for special mention the 27 companies from the *Fortune* list of the 500 largest companies that had accepted the 'climate challenge'. He also recalled a meeting that had taken place shortly before, at Clarence House, the residence of Prince Charles, in which 150 companies had agreed that it was possible to make money from green energy. 'With this perspective,' concluded Nicola Bullard of the Focus on the Global South of Bangkok, 'some companies will win and other will lose, but capitalism will survive.'[43] It was John Kerry who was sent to the Poznan meeting in December 2008 by President Barack Obama.

Then there is the Global Reporting Initiative, which brings together a number of companies round the world for 'a sustainable and transparent development', with four dimensions: ethical (code of conduct), economic (market), social (responsibility) and environmental (sustainability). This is one of the outcomes of the Global Compact, the agreement with the UN that allows companies to use the UN logo in exchange for adopting a code of conduct.

Kofi Annan, the then Secretary-General of the UN, proposed this solution after a decade of failure in negotiations to establish rules for the ecological and social practices of transnational corporations and which would have constituted the beginning of an international economic law, endowed with sanctions and a juridical body. But the opposition of business circles and of the leading governments of the industrialized countries paralysed this project. In contrast, the Global Compact, which relied on voluntary adhesion, enabled the transnational corporations to legitimize themselves by intense publicity campaigns in all the world's media, based on a moralizing discourse, which was very often in contradiction with their practices. This was affirmed by a number of NGOs and by tribunals of opinion, particularly the Permanent Peoples Tribunal in its session at Lima in May 2008, which dealt with the ecological and social practices of the European transnationals in Latin America.

Once again, some people will be tempted to say: what does it matter, in the end, as long as solutions are found? But can one really talk about solutions without formulating the question in a global context? It would indeed be astonishing that an economic system, whatever it may be, does not try to put forward some remedies to a situation that blocks it to the point of fighting for its own survival. But are these solutions favourable for humanity as a whole and prepared to guarantee the future of the planet? In other words, should we save the human species and its vital needs or preserve the future of capitalism? It is indeed an ethical question which is, in fact, theoretically supported by positions taken by the capitalist world.

The first conviction on which neoliberal reasoning is based is the intrinsic nature of the capitalist market as a source of progress. In this perspective, it is inseparable from growth, the principal engine of which is the value of exchange. It is true that capitalism has proved in past history as being the most efficient way of producing goods and services. It also has the advantage of being very flexible and, to the extent that it has the necessary liberty, to adapt itself to all circumstances. It is even capable of transforming the undesirable effects of its own activities into a source of profit and accumulation. According to this reasoning one must therefore conclude that its protagonists are the best placed and often the only capable ones to find genuine solutions to the problems posed by global warming. The role of the public authorities would then consist of creating the conditions for this logic to be exercised, reducing the risks of investment and allowing private interests to take action where profits become possible. In other words, as in many other fields, it is a question of socializing the risks and privatizing the profits, of which the financial crisis is such a clear demonstration.

Such reasoning has the advantage of being clear and is, in itself, logical. It is even capable, once applied, of being efficient in the short and medium term. The reduction of greenhouse gases by industry is an example. The great US financial institutions, Goldman Sachs, Citigroup, Lehman Brothers, have even drawn attention to the advantages of 'green' investment and the shareholders have become aware of it.[44] All of this has also been the subject of intense discussions at the World Economic Forum at Davos, which has become, according to the Japanese Prime Minister Yasuo Fukuda, 'the largest bazaar for relations between investors'.[45]

The only flaw – but it is a very large one – is that the system does not take externalities into account, i.e. the factors that do not directly enter the economic calculations of the market. As in the past, there are still considerable grey zones and we shall see this when we consider the question of agrofuels. As long as the ecological factor remained at the margin of economic construction (an externality) it was not taken into consideration. It was only when the damage wreaked on nature became an obstacle to capital accumulation that the system began to integrate it into its concerns. At the moment, according to its own logic, capitalism will prove capable of transforming conservation measures that have become indispensable, as well as the search for alternatives, into factors of accumulation, i.e. profits.

At first sight this remark would seem to support neoliberal reasoning, as the need to achieve profits (the organic law of capitalism) is the incentive for action. In this the reasoning of Adam Smith combines with that of Bill Gates when, in January 2008, the latter spoke at Davos about 'Capitalism of the Twenty-first Century', emphasizing the system's capacity for self-regulation. But obviously such a position can only be part of the logic of capital accumulation and the first reaction consists of maintaining the rate of profit at all costs. Hence, when the EU announced its wish to reduce the rate of greenhouse gases by 20 per cent, the reaction of certain industries was to plan to delocalize production towards the regions that were less exigent in this respect. The application of this logic therefore is limited to the sectors where accumulation is possible. Worse still, as in the case of agrofuels, it creates new externalities (destruction of nature and social disasters) which will not be taken into account unless, in their turn, they become so negative that the rate of profits is once again halted.

This is the case, for example, of the monocultures of oleaginous palm (the African palm), which is now being planted in tropical zones for, among other purposes, the production of agrodiesel. Pesticides are used that destroy the soil and water supplies and devastate peasant agriculture which, in turn, causes massive displacements of people, unbridled urbanization and external migrations. As long as the gain for capital is greater than the inconvenience and as long as social protest remains controllable, the extension of this agrarian model continues, without taking externalities into consideration. Meanwhile, the damage and the victims are 'the others'. It is at this point that one can see that the capitalist market, which is believed to regulate the economy, is irrational because it is exclusively dominated by only one logic. This led Fidel Castro to reflect: 'Energy is conceived of as another piece of merchandise. As Marx warned, it is not due to individual capitalists being unreasonable or insensitive, but it is the consequence of the logic of the accumulation process which constantly tends to commodify all aspects of social, material and symbolic life.'[46]

This is also to be found in other social sectors because similar logic is applied. For example, as long as the well-being of elderly people is an externality, it remains marginal and is presented as a cost that is often seen as unsupportable for the accumulation process. For this reason resistance to organizing pension schemes has been systematic and they were introduced only after long and tough social struggles. On the other hand, when finance capital could transform pension funds into tools for accumulation policies, including the recourse to speculation, attitudes changed. The privatization of pensions was put on the agenda, exposing those concerned to the risks of commodification when pensions become sources of profit. The beneficiaries had to accept and the price that some of them have had to pay, notably in the US, is well known. This same logic dominates the whole contemporary 'development model', i.e. spectacular growth for some 20 per cent of the world's population, leaving in limbo the 'useless masses', about whom we spoke in our introductory chapter.

The same principle is applied to the new energies and agrofuels: their adoption has to be moulded by the capitalist accumulation process: the concentration of land, monoculture, exploitation of labour, and control of the multinationals over marketing. According to this view, as Eduardo Gudynas has observed, nature necessarily has to be incorporated into the market and managed by economic mechanisms which, in their turn, establish new externalities. Reality is thus reduced to one of its components, which evidently distorts visions and makes it impossible to envisage comprehensive solutions. Karl Polanyi, the Canadian economist of Hungarian origin and a specialist in the history of capitalism, remarked that its characteristic was to 'dis-embed' the economic system out of society and then to impose its own laws on society as a whole.

PROBLEMS WITH THE NEOLIBERAL APPROACH

The main gaps in this approach are very well summarized by Priyadarshi R. Shukla, President of the Indian Institute of Management of Ahmedabad.

> Excessive preoccupation with creating the mitigation regime based on 'rights' (marketable property rights) rather than

responsibility ties (for example the polluter pays); predominant attention on efficiency (cost effectiveness) and little attention to equity in sharing the burden; inability to assess the multiple dividends or penalties of climate change action or inaction; and very little appreciation of the different historical conditions of the developmental state of developing nations in terms of their priorities and capabilities (especially institutional).[47]

In other words, nature has to be incorporated into the market and managed through economic mechanisms, as Eduardo Gudynas has pointed out.[48]

Another aspect of the problem is what about the economies in the South, particularly affected by the question of agrofuels the development of which is presented as being one of the solutions proposed by the market? In fact, the lack of land and the costs of labour oblige the economies of the North to encourage production of agroenergy in the South and, at the same time, in a somewhat contradictory manner, to take conservation measures for the 'carbon sinks' (the forests). In the latter case, the system of the quota exchange of CO_2 and the sale of credits to fix carbon serve to finance the protection of forests as 'machinery to fix greenhouse gas generated by the developed countries'.[49] Thus the dependency of the countries of the South is reinforced (as in the field of raw materials) and at the same time the industrially developed countries can pursue their polluting growth model, even if at a lesser pace. The same author also says that the South is being reduced to playing the role of ecological shock absorber, under the supervision of a transnational green market, which is particularly beneficial for contemporary capitalist production. As for the problems of sharing world wealth or the South's own options for its industrial development, they are completely ignored, while the 'global actors' that are the transnational corporations have a free field.

On the eve of the UN Climate Conference at Bali the US and the EU met in Geneva, to make joint proposals for accelerating the elimination of customs barriers on goods and services, starting with the trade in technologies that reduce the damages caused by greenhouse gases. Victor Menotti of the International Forum on Globalization in Switzerland aptly remarked that such a proposal would be part of the unequal trade framework and that once again it would be the strongest who would obtain the benefits from this kind of transaction. He added that such arrangements inevitably contributed to the weakening of the states in the South, who are less and less capable of taking decisions concerning their own economic destiny. The liberalization of trade inevitably means the domination of the interests of the strongest, even if certain provisional restrictions, in time and space, are envisaged. Trade liberalization therefore works against the well-being of humanity as a whole.

It should also be recalled that the intellectual property rights which, within the framework of the WTO, crown the edifice of international norms, do not benefit the countries of the South. On the contrary they could even become an obstacle in applying and disseminating knowledge about the conservation of the environment. The San Francisco Bay Club organized a meeting with the industrialists of the Silicon Valley concerned with the cleaning up of the environment, at which this fact was recognized and a more open communication about knowledge in this field was proposed. As for the World Bank, it is happy about the proposal of the US and the EU for a liberalization of commercial trading as it believes that it would considerably increase the volume of trade and the value of world commerce. But it never reflects on whom this is going to benefit, nor about the ecological impact of the thousands of cargo ships and lorries required to put all this into practice.

In order to continue this quest for information on agroenergy, it is relevant to ask, what is the situation of energy resources today? We know that the future of fossil energy is calculated in decades, that the utilization of other non-renewable raw materials like uranium is strongly criticized, that research and experiments on renewable sources is booming and that agrofuels are presented by some as an unexpected solution. Thus it is necessary to weigh up the pros and cons of the solutions proposed and to situate them in the framework of the economic logic of the capitalist system, in order to understand whether they constitute a genuine solution to the climate problems or whether they are not more responsive to the needs of the reproduction of capital.

4
Agrofuels and Agroenergy

Like the energy crisis, the climate crisis invites researchers and politicians to find solutions, and among those at present envisaged, agrofuels have taken the lead. The first step is to describe agrofuels and their characteristics, from the agronomic and energy points of view. We then look at the socio-economic context of their production, because there is in fact an abyss between the advantages of their supposed potential and the social and ecological implications of how they are produced.

CHARACTERISTICS OF AGROFUELS

In the same way that hydroelectric power is also known as 'white coal', so one might talk about the 'green coal' energy that comes from plants. These are the agrofuels for vehicles, boilers or cooking stoves. 'Agrofuels are derived from biomass, in principle permanently renewable through the capture of solar radiation, thanks to plant photosynthesis', writes Professor José Walter Bautista Vidal of the University of Brasilia and the father of ethanol in Brazil. He adds that the sun has 11,000 million years of life and each day solar radiation produces the equivalent in energy potential of all the oil reserves in history.[1] One can well understand, therefore, that the countries of the South, which are mostly richly endowed with sunshine and which dispose of huge land areas, are tempted by this solution.

This is also the opinion of Josep Borrell, the then President of the Development Commission in the European Parliament, when he said that agrofuels presented a great opportunity for the South. In Africa, an agrofuel lobby stated that 379 million hectares would be available for this purpose in 15 countries on the continent.[2] In Brazil, according to the InterAmerican Development Bank, there would be some 120 million hectares available. This is why Professor Vidal has said that Brazil could become the supplier of clean and renewable energy for all humanity: an idea that President Lula is putting into practice by increasing the production of ethanol from sugar cane.

75

But Professor Vidal also insists that the use of this source of energy must be compatible with the production of food and respectful of the water table. He proposes the constitution of an international agency for renewable energies to monitor the application of these production conditions.

By definition, in fact, agrofuels are neutral in terms of CO_2 production because, as they are consumed, they send into the atmosphere the quantity of gas carbon that they had captured while they were growing. If one compares the combustion of the motor engine with that of fossil energy, the agrofuels emit fewer greenhouse gases: 60 per cent less for agrodiesel and 70 per cent less for ethanol.

However, this only takes into account the actual combustion process. Apart from the social aspects, which will be discussed later, for a realistic calculation we must include the whole agrofuel cycle, from production to distribution. It could well be that they produce more greenhouse gases than the traditional fuels, if the emissions of an agriculture using fertilizers and chemical herbicides, the manufacturing process and transport are all included. This was what Dr Bernard Pisehesmier, the then President of Volkswagen, had in mind when he said that certain agrofuels were more like 'a wolf disguised as a lamb, because their balance sheet, in terms of CO_2, is even worse than that of the traditional fuels'. He added: 'they are receiving fiscal incentives from limited budgetary resources and are thus a bad investment. One cannot consider them to be sustainable in the ecological or economic sense of the term.'[3]

In spite of this, the production of agrofuels has become a universal preoccupation. In the US, 5 billion litres were produced in 1995, 16.5 billion litres in 2007 and the forecast for 2015 is 56.8 billion litres. According to Richard Greenwald, writing in *Time* magazine on 14 April 2008, the idea of renewable energy has become one of those notions as evident as the Americans' 'motherhood and apple pie', thanks to Richard Branson, George Soros, General Electric and BP, Ford and Shell, Cargill and the Carlyle Group.

We should not, however, forget that, in spite of a large increase in production between 2000 and 2008, agrofuels represent only 1.5 per cent of consumption by transport, 1 per cent of liquid combustibles and 0.4 per cent of the world's energy consumption.[4] A serious contribution to solving the energy contribution would therefore involve a considerable increase in production, above all in the South, with the ecological and social consequences that we know.

THE DIFFERENT TYPES OF AGROENERGY

There are various types of agroenergy. The most important one is ethanol (as a substitute for petrol) which is an alcohol produced through the fermentation of simple sugars (such as beetroot or sugar cane) that come either from plants rich in starch (such as potatoes and cereals) or from ligneous plants (such as trees or straw). It is also possible to produce an ether (a product of the reaction between an alcohol and an acid) derived from ethanol, the ethyl tertio-butyl ether (ETBE).

Another type is an ester (a chemical composite produced by the reaction between an alcohol and an oil) of plant oil or agrodiesel (substitute for diesel oil).[5] To help us understand these processes, we shall now examine the plant sources of the first-generation and second-generation agrofuels.[6] (According to some experts there are actually three generations, depending on how one classifies them, but we shall stick to the more usual definition that uses two.)

First-generation Agrofuels

The so-called first-generation agrofuels, a list of which will be given later on, are the products of alcohol (ethanol) or vegetable oil (agrodiesel), intended to become the equivalent of fossil fuels, petrol and diesel, respectively. They usually originate from cereals or plants already used for feeding humans and animals, or for industrial usage (pharmaceuticals and cosmetics). Ethanol is used much more than the vegetable oil methyl esters (VOME) or agrodiesel, the consumption of the latter being about one-tenth of the former.

While ethanol is mainly produced in the US and Brazil, agrodiesel is still a specifically European product. The US, Brazil and Europe thus predominate in the production and consumption of agrofuels in the world. Their production has increased considerably over the last few years, particularly since 2002, and it is predicted to increase still more in the future. In fact, since this date the annual increase of the world production of agrofuels has been about 15 per cent. There are now numerous developing countries that are launching huge programmes for the production of agrofuels from sugar cane or succulent plants such as the oil palm and the jatropha. At the same time the European Union is reducing its 'set aside' programmes.

Let us look a little more closely at production from alcohol and ethanol.

Agrofuel from alcohol: ethanol

The alcohol-producing plants mainly used are beetroot, sugar cane, maize, wheat, barley, potatoes, Jerusalem artichokes and sugar-producing sorghum. In the US it is maize (which they call corn) that is used on a large scale to produce ethanol, but with a much lower yield than sugar cane, which has mainly been used in Brazil, particularly since the 1960s.

The other plants are much more marginal. Cane from Provence, France, has an average production of some 20 tonnes of dry matter per hectare per year and is also used to produce thermal energy. The same is true of hemp, melilot (sweet clover) and the freshwater hyacinth. Certain prairie plants could be specially developed for energy use and studies are being carried out in this field. Thus energy can be produced from the stalks of lucerne which has protein-rich leaves, although the drying remains a problem because this cannot be done while still on the stalk.[7]

The freshwater hyacinth has certain advantages. It grows best at a temperature between 25 and 30 °C, so it is being studied in stretches of warm water, particularly in water discharges from thermal electricity plants. In Ile de France, under glass and in warm water, it produces 140 to 230 kilograms of biomass per hectare each day. It is fed by urban or agricultural liquid industrial waste and thus combines decontamination with energy production.

Agrofuel from oil: agrodiesel

Fuel-oil yielding plants, also known as pure plant oils (PPO) or crude vegetable oils can be used (up to 100 per cent) as a fuel for all diesel engines (which were invented for this kind of fuel), but they need minor modifications to heat the fuel or, without modification, to be mixed with ordinary diesel (30 per cent for all vehicles and up to 50 per cent for some of them). But oil is also the main crude material used to make agrodiesel (strictly speaking), which is an alcohol ester used as fuel incorporated into diesel.[8]

Agrodiesel is thus the second plant fuel used in the world after ethanol, but its contribution is still modest, with global production estimated at 3.7 million tonnes a year, which represents barely 10 per cent of the total ethanol production. Its consumption is assured in the future, especially in Europe, as there is a large increase in diesel-powered vehicles: about two-thirds of newly registered vehicles in Europe have a diesel engine.

Combustion engines can use both vegetable oil (colza, sunflower, palm, soya, peanut, etc.) and oil esters.[9] Esters have two advantages over crude oils: they have less viscosity and a greater capacity to self-ignite in the motor of a vehicle. Some agricultural tractor makers are producing engines that can use oils that have not been esterified, but the fuel most used in Europe today is the methyl ester of colza oil. Trials carried out with 30 per cent ester in municipal transport vehicles in several dozen towns have proved that this creates no problem for the engines.

In 2003 and 2004 Daimler-Chrysler, in collaboration with the Indian Central Salt and Marine Chemicals Research Institute and with the University of Hohenheim in Germany, tested agrodiesel obtained from the oil of *Jatropha curcas* seeds on three specially adapted Mercedes. In 2005, these cars travelled 30,000 kilometres under difficult conditions, crossing over mountain passes of over 5,000 metres above sea level, without any problems.[10] Nevertheless, Professor Rudolf Maly, chief of the Daimler-Chrysler project, points out that this fuel has still not reached its optimal quality. But it already satisfies the European norm and is simple to produce.

The raw materials used to obtain agrodiesel are numerous oleaginous plants with oil yields that vary from one species to another. They include green algae, almonds, peanuts, colza, olives, the African palm (nuts and kernels), raisin pips, castor oil, sesame, sunflowers, mustard, soya, manioc, canola, buriti palms, and proteaginous peas.

The euphorbe family (mainly the *Jatropha curcas*) which can be developed in dry and poor conditions, has a latex (a thick milky juice) from which it is possible to extract hydrocarbons and its seeds are rich in oil. Finally, there are the shrubs like gorse and broom, which adapt well to poor soils and difficult climatic conditions and can produce quite high yields. However, because of the enormous current and future demand for energy, most of these sources cannot compete with fossil fuels.

Clearly the non-food crops are the best alternative for producing agrofuels as they make it possible to limit the use of land allocated for this purpose. Other plants can be used besides *Jatropha curcas*, such as *Moringa oleifera* (drumstick tree), *Pongamia pinnata* (or karanj), *Madhuca longifolia* (mahua), *Cleome viscosa*, flax, copra, eucalyptus and the butter tree.

Second-generation Agrofuels

In order to mitigate the effects of using food crops for producing fuel, which has been criticized around the world, research has been using other sources, for example transforming lignin and cellulose from plants (straw, wood, waste) instead of sugar and starch or using micro-algae from the sea. The latter process makes it possible to obtain yields in oil that are 30–100 times higher than using land plants. There are known to be more than 100,000 species of micro-algae in the world and each year almost 400 new taxons are discovered. Apart from not using food crops, the second-generation agrofuels have several advantages over the first because they require fewer fossil inputs and also aim at using the whole plant,[11] which is the object of much current research, particularly using pyrolysis and carbonization processes.

The fuel potential of wood and ligno-cellulose is receiving special attention. The idea is to develop the production of fast-growing trees and use the woody matter to produce fuel. This requires new techniques that have not yet been perfected, because it is not enough only to transform the biomass into alcohol or to extract the oil of certain plants, but also to use the trunk and branches themselves, hard materials that have to be crushed in order to transform them. Up until now, plantations of eucalyptus, poplars or pines serve above all to produce pulp or are transformed into charcoal. However, it is envisaged that they will be transformed into fuel. In order to accelerate the process of growth and hence productivity, there are trials going on to produce GM wood.

Wood has not generally been included in the category of agrofuels, which relate more to liquid products. In fact, over the ages, wood has always been used as a solid combustible. For thousands of years it has even been the only one serving domestic and industrial purposes. Sometimes it is used directly, sometimes as charcoal. In the continents in the South, it is mainly for cooking and heating that forests are being exploited (more than 75 per cent of the wood is used for energy and less than 25 per cent for timber; in contrast, these proportions are inversed in the industrialized countries).

Since the 1960s, plantations have appeared alongside natural forests. They are aimed specifically at producing maximum energy (and not necessarily for furniture): eucalyptus in Brazil, poplars and willow in Europe. Better genetic selection of the tree species and new methods of cultivation and harvesting have improved yields: for example in Europe, felling willows and poplars after five to seven years makes it possible to produce 10–13 tonnes of dry wood per hectare (as against 3–5 tonnes for a classic forest). But wood energy accentuates atmospheric pollution. It is true that wood has practically no sulphur and its combustion therefore does not emit sulphurous SO_2, but it discharges many particles in its smoke. It also emits hydrocarbons and organic composites that condense into liquid tars and we do not yet have any reliable statistical information. A more complete combustion, at a higher temperature, would however make it possible to diminish these undesirable discharges.

The other gases emitted by wood combustion (CO_2, azote oxide – NOx, methane – CH_4) vary considerably, according to the combustion equipment. But progress is being made. Thus, in Austria, the 'kingdom of wood heating', the emissions of pollutants by heating equipment are now ten times fewer in as many years. As for the CO_2 discharged by the combustion of wood, it is reabsorbed by plants and trees for their growth and hence recycled. From this viewpoint, the wood sector, for the same quantity of energy produced, contributes 12–15 times less greenhouse gas than coal and 7–12 times less than fuel oil or natural gas.[12]

Nevertheless, as well as the atmospheric pollution, it is also necessary to mention the chemical pollution due to forest exploitation and deforestation, as a result of using wood as a source of energy. Other factors also have to be taken into consideration: the drying up of the soil by the absorption of large quantities of water and the use of fertilizers and chemical pesticides, not to speak of the effects of monoculture, which we shall discuss later. In fact, methanol or 'wood alcohol' obtained from methane by wood transformation is a fuel that can partially replace petrol, or can perhaps be used as an additive to diesel oil and in certain cell fuels. Cellulose, considered one of the most universal molecules, can be transformed, thanks to enzymatic degradation or gasification, into alcohol or gas serving as agrofuels. This is beginning to be implemented in Canada, the US and Sweden, but its perfection will take several years yet and the ecological and social effects of its production are not sufficiently known.

5
Ethanol Production

FROM SUGAR CANE

It is not enough to know the characteristics of the various kinds of agrofuels and what they can theoretically offer as solutions to the climate and energy crises. They must be put in their concrete context, that is to say their production must be analysed as well as the ecological, economic and social consequences of their transformation and distribution. So in this chapter we shall study some of these examples, both of the ethanol sector in Brazil and, in the next chapter, in the agrodiesel sector in Colombia, Indonesia, Malaysia and Africa.

Remember that ethanol is the result of transforming sugar or starch into alcohol, which can be used directly as fuel, or it can be mixed with petrol. Ethanol emits 70–75 per cent less CO_2 during engine combustion. But its real effectiveness compared with fossil fuel can be disputed if one takes into account the whole cycle of production and distribution. An article in *Science* magazine states that if one considers the deforestation that it causes, ethanol from maize and diesel from soya double the production of greenhouse gases. According to Professor David Tillman of the University of Minnesota, 93 years must pass before ethanol recovers the carbon emitted by the clearing of the land used to produce it.

IN BRAZIL

Brazil is a particularly good example of ethanol production because it is second only to the United States among the largest world producers. The oil crisis at the beginning of the 1970s encouraged the country, with its huge extensions of sugar cane, to use this source of energy. The return of cheaper petrol ended this first craze, to the point that the World Bank and the IMF put pressure on the government to lift its subsidies to agrofuels. The national oil company Petrobras also was little inclined to encourage the new sector. But with the new oil crisis and the price explosion, production of agrofuels has started up again. Since 2004 half the number of Brazilian cars run on alcohol, pure or mixed, and in 2007 the proportion had reached 80 per cent. That same year, according to the Menerval Fuel Association, 19 billion litres had been produced and the forecast for 2010 is 70 billion.

The aim is to attain a production of 100 billion litres a year, using 30 million hectares of land – five times more than that cultivated in 2007. Such a figure is theoretically possible, as Amazonia alone, according to its advocates, can supply 70 million hectares. There are thus reserves as far as land is concerned. That is the view of the engineer Expedito Parenti, and he goes even further: 'We have 80 million hectares in Amazonia which will become the Saudi Arabia of agrofuels. In fact it is not only sugar cane but other crops, such as sunflower and soya, could also take up some 60 million hectares.'[1] And we should not forget that some 42 million hectares are in illegal possession (*em grilagem*).

To encourage this process, Law 693 of 2001 envisages the use of 10 per cent of ethanol in fuel consumption for 2009 and, if possible, 25 per cent towards 2025. Sugar cane cultivation has increased rapidly: in 2007–08, 6.6 million hectares were given over to it – 7.4 per cent more than the preceding year. Some 528 million tonnes of sugar cane were produced, of which more than 88 per cent was destined for ethanol.[2] Between now and 2014, there are plans for 114 factories for processing sugar cane. In 2005, 2.5 billion litres were exported to the US, Japan and Sweden, and these figures will surely increase.

As can be seen, Brazil is heavily involved in the production of ethanol. In 2006 an agreement was made with the US, a country that is particularly concerned to reduce their dependence on fossil fuels from the Middle East or from countries felt to be unreliable like Venezuela. In 2005 the US imported 58 per cent of its consumption of ethanol from Brazil and if it is to attain the objectives fixed by then President George W. Bush for 2017, it must procure another 135 billion litres of Brazilian ethanol per year. The US produces ethanol from maize, which yields 3,037 litres per hectare, but in Brazil one hectare of sugar cane produces as much as 6,879 litres.[3] There is talk of an OPEC of agrofuels. Petrobras is also associated with the project. In 2007 President Lula made a tour of Europe and established contacts with the EC to present the advantages of his energy policy.

At the time of the European-Latin American Summit in Lima, in 2008, certain doubts about agrofuels were raised by Europeans but the Brazilian delegation was extremely anxious to defend their position. President Lula wants to reach an agreement to re-establish the negotiations of the Doha Round within the WTO. He stated that Brazilian policy on this matter wants to contribute to the well-being of humanity. In July 2008 he signed an agreement with President Uribe of Colombia for developing agrofuels and their common declaration affirms that agrofuels do not affect the price of food products. Both used very harsh words in criticizing the movements and organizations that opposed their projects.

However, it is important that the arguments not be confined only to considerations of productivity per hectare and improvement in the conditions of combustion of agrofuels. Consideration must also be given to the ecological and social effects of their production and the type of economic model that defines the context.

Economic and Social Dimensions of the Brazilian Model

As far as the environment is concerned, the effects do not differ from those that we already mentioned *apropos* of monoculture: the use of fertilizers and pesticides that are dangerous for biodiversity, for the quality of the soil and water, and for the health of human beings. In the region of São Paulo, where there is a large area of land given over to sugar cane, the acidity of the soil has greatly increased, which tends to make other crops disappear, for example fruit cultivation. In the same region, in order to clear the land, 60 per cent of the cane trash has been burnt, which is particularly harmful for the environment. Micro-organisms in the soil are destroyed and the air polluted which causes respiratory diseases. It also creates a drop in the degree of humidity, from 13 to 15 per cent according to the Brazilian National Institute for Space Research.

Sugar cane production in Brazil does not directly encroach on forest land, particularly the Amazonian forest, which is not a sugar cane producing area. In fact, in a number of the states that are now producing cane sugar, their original forests were destroyed long ago. Nevertheless, indirectly, the current extension of sugar cane results in displacing pastureland and soya, particularly towards regions that are now wooded, especially in Amazonia. The destruction of small-scale agriculture by land concentration also chases out the peasants, some of whom become legal or illegal colonizers of forest areas while others migrate to urban slums. The law passed on 4 August 2008, which increased the amount of territory which could be sold without submission, may very well promote this trend.

Cerrado, in the centre-north of the country, is particularly vulnerable. Its livestock production is being displaced by eucalyptus plantations, but also by sugar cane, which in 2007 occupied 5.5 million hectares. This is one of the richest biodiversity areas in Brazil. Some 10,000 plants have been registered, many of them unique in the continent and the number of mammal species exceeds that of Africa. Cerrado has lost half its land surface in 40 years and in the country as a whole 162,000 hectares of what is called 'the conservation zone' have already been taken over by sugar cane production.

As a consequence of monoculture extension and hence agrofuels, the population has been displaced. All over the country (and evidently for reasons that are not only due to this agricultural sector) between 1985 and 1996 – just ten years – 5.3 million people were uprooted from their land, representing the disappearance of 940,000 peasant holdings.

The growth in agrofuel production in order to increase the country's revenues underpins the logic of governmental decisions. According to this perspective it is a question of facilitating a redistribution of wealth, particularly through the programmes Zero Hunger and Family Assistance which have already proved themselves to be efficiently managed and capable of reducing destitution and hunger.

Plans for the production of agrofuels in Brazil are drawn up for the short and medium term, based on a lot of research. For the long term, the aim is to produce cellulosic ethanol, a second-generation agrofuel, which could have results towards 2015.[4] This can only encourage eucalyptus monoculture, for example, with all its long-term consequences of the drying up of soils, as well as the development of GM plants to increase productivity. Hence, in the first three months of 2007, 6.5 billion dollars were invested in this sector – 66 per cent more than during the same period in 2006.

The source of these investments come from both Brazil and abroad. In the former case, big companies are involved, like Odebrecht, specializing in petrochemicals, which has decided to invest 5.3 billion dollars in the production of ethanol between now and 2013. In ten years, the company hopes to produce 30–40 million tonnes. Other large companies are following suit, both in the transformation and distribution of agrofuels, like Cosan, Bonfim, CDC, Bioenergia, Guarani and, of course, Petrobras.

Foreign investment is also necessary to achieve production targets. Not only are large corporations like Cargill, Bunge, ADM and Syngenta buying huge extensions of land for sugar cane monoculture (or soya and palm oil for agrodiesel), but financial capital from the US and Japan is also interested in the sector. George Soros has decided to invest 200 million dollars for producing alcohol in Minas Gerais and Bill Gates is contributing 86 million dollars to finance the company Pacific Ethanol, in order to guarantee supplies for the US. Similar steps are being taken by James Wolfensohn, the former director of the World Bank, and Vinod Khosla of Sun Microsystems. As for the Japanese, an agreement has been signed between Petrobras and the Nippon Alcool Banki, to create the Japanese Ethanol Co. The Sumitomo Corporation, Mitsui and the Japan Bank for International Cooperation are also active in this field, not to mention the Europeans, especially the Swedes.

The economic model is clearly oriented towards exportation, which will take up three-quarters of Brazil's production over the coming years in order to supply 50 per cent of the world market. But there are obstacles. The country's infrastructures are insufficient, in terms of roads, river ports and means of transport. This could well constitute a brake on export plans. It is true that an alcoduct 1,150 kilometres long is envisaged between the region of Goyas and São Paulo. It will be capable of transporting 6 billion litres of ethanol a year, which would enable the production of Goyas to be doubled between now and 2013 and it will be 16 times less expensive than road transport. It will need an investment of 500 million reals (more than 200 million dollars).

The model based on monoculture also has social consequences. It assumes a considerable elimination of labour, particularly of the small peasants. In 2005, there was a loss of 300,000 jobs in agriculture. This increases internal migration, uncontrolled urbanization and pressure on the agriculture frontier.

Furthermore, work in the sugar cane plantations is particularly hard. The working day is long and the pace intensive. In El Salvador they cut between 5 and 12 tonnes a day (in Brazil the figures are often even higher) and they work seven days a week for a wage equivalent to 2.50 dollars a day. According to a study carried out by Fontana de Laat and published by the Movement of Landless Peasants (MST), in 2008 the cane cutters, every ten minutes, cut 400 kilograms of cane, with 131 blows of the machete, requiring 138 body movements. This results in heart fatigue. In one day, it amounts to the cutting of 11.54 tonnes of cane, 3,792 blows of the machete and 3,994 body movements. Breaks, which are allowed every 30 minutes, are mostly not respected, so there are serious health risks and the life expectation of workers is badly affected. On top of this the wages are too low, at the limit of subsistence, which means that we are really talking about a new form of slavery, and child labour.

There is no doubt that such exploitation is responsible for the large profits for agro-exporter landowners and for the national and foreign companies. It all reinforces an unequal social structure, which was already one of the worst in the world and it is not consistent with the projects to reduce the gap promised by the Workers Party (Partido do Trabalho).

At the political level, the North/South dependency structure is reinforced. Indeed, it builds an integration into the internationally dominant economy, which contradicts the efforts now being made to bring the Latin American countries together, particularly in the framework of the Bolivarian initiative, ALBA (Bolivarian Alliance for the Peoples of our America). But it is in line with President Lula's vision: strong economic growth which makes it possible to free up the means for a social policy that aids the poorest of the poor. It also poses the question of the basic philosophy of this policy – i.e. its ecological and social cost and the absence of the structural reforms necessary to enable the less privileged groups to become genuine actors in society, while allowing the continual reproduction of the gap between the rich and the poor, even in a somewhat mitigated form.

It cannot be said that the Brazilian government has been indifferent to the problem. João Pedro Stedile, founder of the MST, who is an economist, published an article on the issue in the *Monthly Review* of February 2007. He said that the measures that had been taken by the Lula government to promote peasant agriculture were remarkable. He particularly mentioned better access to credit and insurance, a big effort to bring electricity to rural areas, housing construction, increased technical assistance, the delimitation of indigenous territories and less political (federal) repression.

On the other hand, he wrote, the macro-economic policies promote agricultural trade, above all for international commerce and are inspired by the neoliberal policies of the WTO and the World Bank. They oppose, for example, the labelling of transgenetic products. Lula's government has maintained tax exemption for exported agricultural products and legalized transgenetic soya, while public banks have increased their support for agrobusiness (12 billion dollars for the 2006–07 harvest, of which 4 billion for the largest transnational agro-industries). It should be added that a number of electoral promises have not been fulfilled: particularly a genuine agrarian reform, a review of the productivity index, the expropriation of plantations using slave labour, the control of soya and cotton monocultures and the creation of agro-industrial cooperatives for peasants.

However, an alternative model based on peasant initiative does exist in Brazil. The best known one is the Bindozana cooperative at Alagoas, described by Ignacy Sachs in a document titled *Biocombustiveis os alimentos concurrência o complementaridade* ('Bio-fuels: competing or complementing food?').[5] There is also the Cooperbo, a cooperative organized jointly by the MST and the Small Peasants Movement (MPA) in Rio Grande do Sul. The social movements and the Pastoral Land Commission of the Brazilian Bishops Conference (CNBB) strongly advocate that peasant agriculture be given priority. In 2007 a conference on agrofuels in Curitiba brought together various social movements, experts in this sector and strong defenders of agroenergy who saw it as a question of Brazilian pride and nationalism. But they were all concerned about social justice.

The conference resulted in a declaration dated 31 October 2007 entitled 'For Food and Energy Sovereignty'. It stated the need for a harmonious relationship between humanity and nature, which means respect for biodiversity, soils and water, and excludes monoculture as well as extending the agricultural frontier. The production of energy cannot take the place of food, or be determined by the laws of the market. The declaration also demanded an agrarian reform and said that the initiatives in the field of agrofuels must first respond to local and regional needs, more than to exportation. Also production must be decentralized, on the basis of peasant agriculture. As can be seen, it is not a question of an absolute rejection of agrofuels but rather a list of the ecological, economic and social conditions of their production and control.

The conclusion of João Pedro Stedile is that the Lula government is ambiguous, for while the ministries of agrarian reform and of the environment defend the family model of agriculture, those of the economy, industry, trade and agriculture promote agricultural trade. The latter model has benefited from the power of its promoters, while agrarian reform has been virtually paralysed or reduced to measures of social compensation.

IN OTHER COUNTRIES OF THE SOUTH

Other countries are following suit. In the Caribbean it is Jamaica, and in Central America, Guatemala, Honduras and El Salvador. Workers from Honduras and Nicaragua come to work as cane cutters in El Salvador. Agreements between Presidents Bush and Lula have been especially concerned with the development of refineries in El Salvador, in connection with the plantations of Honduras and Nicaragua, thus reinforcing the power of local capital (the Pellas in Nicaragua, the Maduro in Honduras, the Calderon in El Salvador). Elsewhere in Latin America the production of ethanol is booming, for example in Ecuador where more than 50,000 hectares are dedicated to it, with assistance from China. In Mexico, given the importance of maize in local diets, a law was adopted in December 2002 limiting the production of ethanol to surpluses of white maize.[6] In Venezuela, however, the effort being made to increase the production of maize according to the 2007 plan, excludes transformation into ethanol.[7]

Brazil's action has spread to Africa, where some 15 countries have made agreements for the use of the Brazilian technology: Benin, Burkina Faso, Cape Verde, Côte d'Ivoire, Gambia, Ghana, Guinea, Guinea Bissau, Liberia, Mali, Niger, Nigeria, Senegal, Sierra Leone and Togo. In November 2008 an international conference was organized by President Lula, who announced that there would be an increase of 200 per cent in agrofuels between now and 2014. Some 2,000 people attended the conference, including nearly 80 ministers. President Lula made the Brazilian Trade and Investment Promotion Agency (APEX-Brasil) responsible for organizing the first international exhibition on biofuels.

In Asia, the traditional sugar cane growers are increasing their yields to get into the market for ethanol. For example, in the Philippines, particularly the island of Negros, the plantations are encroaching on state lands.[8] However, the food crisis is putting a serious halt to this expansion (the Philippines has to import several million tonnes of rice each year). In Hawaii, the local government adopted a law in 2006 requiring consumption of 20 per cent agrofuels by 2020. There, too, foreign investors are interested, including Vinod Khosla, the owner of Sun Microsystems who has invested in Hawaii Bioenergy. As for the three great landowners who possess 10 per cent of the island's territory, they are also moving into this field.

IN THE COUNTRIES OF THE NORTH

The South does not have a monopoly over ethanol. According to Dominique Parizel, in the EU the production of ethanol in 2005 was 0.73 million tonnes, compared with 26.9 million at the world level.[9] In Italy there are projects to produce it from maize and in Belgium it is encouraged by federal and regional governments, mainly from the production of sugar beet. The factory of Wanze, in Wallonia, which is a dependent of the German Südzüdcker companies, is supposed to produce 300 million litres a year. For this purpose it will use 800,000 tonnes of wheat and 400,000 tonnes of sugar beet. In Flanders those concerned are Acco in Gent and Amylum in Aalst. As production is not profitable there has to be some state participation, particularly through tax deductions for the sector. A further problem is that there is not enough land in Europe to accommodate the demand for agrodiesel, which means a massive recourse to land in the countries of the South.

In the US, 15 per cent of the land available was used in 2007 for agro-fuels (it requires 121 per cent to meet the needs defined by policy). Hence the desire to maximize production and to use GM plants – what some American farmers call 'Monsanto moonshine'. In Louisiana, where the coastal waters are polluted by nitrates, the National Academy of Sciences has raised the alarm. However, the US Senate envisages the production of agrofuels to increase from 28 billion litres to 136 billion by 2022, with 57 billion being produced from maize starch. The quantity of nitrate carried by the Mississippi into the Gulf of Mexico will increase during this period from 10 per cent to 34 per cent, which will turn it into a veritable bin of toxic rubbish from the Green Belt. This is what is causing the famous 'dead zone' which already affects many coastlines, including those of Brazil. In the summer of 2007 it was calculated that this stretch of water, where there is no maritime life except for algae, covered 20,000 square kilometres in the Gulf of Mexico.[10]

6
Agrodiesel Production

FROM PALM OIL

ORIGINS AND CHARACTERISTICS

Palm oil can be used in many ways. First of all it is a basic product in food, for margarine, table oil, dairy ice cream, chocolate, precooked meals, animal feed, etc., but it is also much used in the production of paints and varnishes. The pharmaceutical industry, too, is a strong consumer. And there are also a couple of dozen interesting sub-products that use it, such as furfural, a natural bactericide, fungicide and insecticide, as well as the lignin that comes from the wood of the palm tree and serves to make plywood.

In tropical Africa, palm oil has been used for a very long time by the population. Small quantities have been exported ever since 1583. As from the end of the 18th century, it became a veritable export product, replacing the slave trade.[1] In 1840, England obtained 15,000 tonnes of palm oil. In 1911, in the Belgian Congo, Lever Brothers created plantations and factories, thus moving from the harvesting of wild palm trees to industrial production. In 1913, France imported 200,000 tonnes of oil which came from the mesocarp (the fleshy part of the palm nut) and 300,000 tonnes of oil derived from the kernel, but the introduction of the plantation system diminished production of the latter.[2]

It is only recently that palm oil has been transformed on a large scale into agrodiesel, especially since oil prices have shot up. While European demand for products derived from palm oil has remained more or less stable since the 1980s, that of India, Pakistan, China and the Middle East has exploded. This new market, like that of Eastern Europe, developed all the more rapidly because part of the population had adopted Western consumption habits.[3] Hence the enormous extension of palm cultivation in all the tropical and subtropical parts of the world. At the beginning of the twenty-first century it covered 20 million hectares of land.

The annual yield of oil palm trees is 5,000 tonnes per hectare. But the sector is only profitable today in countries where labour is very cheap.

CONTEMPORARY PRODUCTION

In Asia (Mainly Malaysia and Indonesia)

Cultivation of the oil palm has been going on for decades in Malaysia, since the end of the 1950s when rubber plantations were transformed into oil palm cultivation under the auspices of the Federal Land Development Authority (FELDA), which, in 2005, was responsible for 20 per cent of national production.[4] Today, writes the *Malaysian Star* confidently, 'The demand for agrofuels will come from the European Union. This new demand should, at a minimum, absorb most of the palm oil stocks of Malaysia.' In fact, the government is preparing a national agrofuel policy to encourage oil palm production as well as its internal consumption. Some 54 projects for producing B100, an agrodiesel that is 100 per cent palm oil, have been approved by the government, which has joined with private partners to construct three production factories of the new fuel, destined for exportation.

Women supply 50 per cent of the labour as temporary workers in Malaysia, spreading toxic fertilizers and pesticides on the plants. There are many accidents and illnesses among both men and women. The leaves of the palm trees are very sharp and the work is dangerous for the eyes (in the same way that the stems of sugar cane harm many of the Brazilian workers). There are very few medical checks. Frequently there are reports of skin diseases, genital burns, fatigue and headaches, as a result of the chemical products and long working days without any rest.

In Indonesia, almost one-third of the palm oil is produced by small farmers who have often lost their right to land as the plantations expand. Having two hectares as 'retribution', they find themselves tied hand and foot to the palm oil industry, which gives them credits in exchange for their harvest. It goes without saying that they do not receive the best price for their production. According to Abet Nego Tarigan, Deputy Director of Sawit Watch, an organization that represents the interests of the rural communities, farmers and agricultural labourers affected by the production of palm oil in Indonesia, the decisions taken in Europe concerning agro-fuels have direct consequences for millions of people in Indonesia. In their mad rush, the powerful palm oil producers have no hesitation in expelling communities from land that they have been cultivating for many generations. The waged labourers and the small farmers are exploited unscrupulously and we are going to lose very valuable agricultural land, which we have been cultivating to produce food that we need to eat and earn our living. The projects proposed by the European Union will make the situation still worse. If nothing is changed, the poor will be more and more numerous and all the land will end up in the hands of a few.[5]

Today there are nearly 6 million hectares under palm oil cultivation in Indonesia, and three times this area, some 18 million hectares of forest, have been cleared for its expansion. Regional plans envisage dedicating a further 20 million hectares for this purpose. Projects are being discussed that would set up, in the heart of Borneo, a palm oil plantation of nearly 1.8 million hectares, which will be the largest in the world. In three years alone, almost 1.3 million hectares of forest have been destroyed. These plans and projections will have major effects on the remaining Indonesian forests and on the populations that depend upon them. The country has already lost 72 per cent of its ancient forests and 40 per cent of all its forests.

The demand for land has caused a new surge of deforestation. To accelerate the process in Indonesia, forests are being destroyed by fire. In 1997 and 1998 there were gigantic firestorms over a region greater than the Netherlands. It spread smoke that reached Thailand and the Philippines and discharged thousands of tonnes of CO_2 into the atmosphere. George Monbiot, in an article in the *Guardian* of 8 December 2005, drew attention to the massive destruction that was going on in South-East Asia to provide agrofuels for the rest of the world. He argued that by promoting vegetable fuels as is done by the EU, the US and thousands of ecologists in the belief that it creates a market for used frying oil or colza oil, in reality a market has been created for the most destructive cultures of the planet. Nevertheless, four new refineries are being constructed on the Malaysian peninsula, one in Sarawak and two in Rotterdam. Two foreign consortia – one German, one American – are setting up rival factories in Singapore. All these companies are going to be making agrofuels from the same source: palm oil.

Everyone should see what is actually going on in the field. Vast forests of trees, containing large stocks of carbon, are cut down and burnt before the planting of the oil palm trees, which are small and stunted. The driest areas are used first, then the plantations move towards the marshy land of forests which grow on peat bog. After cutting down the trees, the planters dry the soil. When the peat bog is dry it oxidizes and gives out more carbon than that contained in the trees. In terms of the impact on the environment, local as well as international, palm oil as an agrofuel is still more destructive than crude oil from Nigeria, claims George Monbiot. The British government has understood this very well: in a report dated August 2008, it was stated that the main environmental risks will probably be linked to a large increase in the production of raw material for agrofuels, particularly in Brazil from sugar cane and in South-East Asia from the oil palm plantations.

The development of agrofuels in Indonesia has attracted investments from European, Japanese, Chinese and American transnational corporations and the area of land cultivated for producing palm oil should increase between now and 2020 to an area equivalent to almost five times that of the Netherlands. According to Sawit Watch, which is a partner of Oxfam, some 400 communities will be involved in disputes concerning land destined for palm oil cultivation. The largest expansion of these plantations is in the region of West Kalimantan.

Both in Indonesia and Malaysia this development is also producing another ecological disaster. Apart from the deforestation and the elimination of biodiversity, for each harvest to get the 5 tonnes of oil produced on a hectare, there will be no fewer than 40 tonnes of solid waste. These are often burnt, thus discharging large quantities of CO_2 in the atmosphere.

Moreover, there have been numerous irregularities in the way in which the palm oil companies acquire and conserve the land, such as not recognizing customary rights (often the plantations are established without governmental permission), the absence of any information for the local communities, agreements that have not been negotiated, traditional leaders manipulated to force the sale of land, indemnities not paid, advantages promised but not fulfilled, land not attributed to the small farmers or not developed, peasants burdened with unjustified debt, studies on the environmental impact carried out too late, the land not developed within the planned time frame, the use of coercion and violence to crush community resistance and serious violations of human rights.

The problem is that it is difficult to produce oil palm trees integrated with other crops as they are voluminous and their fibrous roots spread out very far. Each palm tree weighs more than 3 tonnes and there are few other plants that can be included in the plantations. For the creatures that live in the soil, such as earthworms, they find it difficult to make paths for themselves. And it is also very difficult and costly to get rid of dead palm trees and their roots, because a backhoe (retrocaveuse) is necessary to dig up the roots, or else chemical products have to be used to destroy them. The FAO is already alarmed by the effects on food security and many environmental organizations are posing the problem of ecological disturbances.

According to the World Bank, the changes in the utilization of the soil, such as deforestation or the drying of bog peat to produce, for example, palm oil, could cancel out the advantages of reducing greenhouse gases for decades.[6] The Bank proposes setting up certification systems making it possible to measure and indicate the environmental performances of agrofuels (for example a green index of reductions in greenhouse gas emissions) which could contribute in reducing the environmental risks associated with the production of agrofuels on a massive scale. However, to be effective, such measures would need the voluntary participation of all the large producers and buyers, as well as creating efficient control arrangements. This is far from being put into practice now, as it enters into conflict with the immediate interests of the producing companies and the governments of the countries concerned.

Papua New Guinea is another large producer of palm oil, reaching an annual amount of 380,000 tonnes, but this is way below the production levels of Malaysia and Indonesia.

Other Asian countries are now also involved. In the Philippines and Thailand 6 million hectares were planted with oil palms in 2007; by that year Thailand was already producing 8.5 million litres of palm oil from 400,000 hectares. and the aim is to double the land planted with oil palm by 2009.

Cambodia, India and the Solomon Islands are also producers. In China, the consumption of vegetable oil has doubled between 1996 and 2006, reaching 24 million tonnes in 2007, with the prospect of increasing by 500,000–600,000 tonnes a year. Palm oil provided 24 per cent of the total in 2000. The country also imported nearly 5.1 million tonnes in 2006–07.[7]

In Africa

Africa provided most of the production until the 1960s, which is why the Latin Americans call it the African palm. At that time, the continent supplied 74 per cent of oil from the palm nut and 50 per cent of oil from the kernel. In 1989 the proportions had been reduced to 14 per cent from the palm nut as against 78 per cent for Asia, and 21 per cent for oil from the palm kernel. The trend has grown even faster, and in 2000–01, out of a world total of 23,361 million tonnes of palm oil, Malaysia and Indonesia produced 82.6 per cent, Africa 6.5 per cent, and South America about 5 per cent.

Nevertheless, today there is increasing interest in Africa in developing the cultivation of oil palm. Nigeria had 160,000 hectares of palm trees planted in 2003. The figure rose to 300,000 in 2007, and the country is getting ready to dedicate 3 million hectares for this purpose in the near future. In 2007, in Côte d'Ivoire, 250,000 tonnes of palm oil was produced. In the Democratic Republic of Congo, 214,000 hectares were planted, and this figure is soon to be increased, with the help of European, Japanese, Chinese and American investments, to 3 million hectares in the Equatorial, Bandundu and West Kasaï provinces. By 2007 Cameroon was producing 250,000 tonnes of palm oil a year, with assistance from France, the World Bank and the IMF.

In Latin America

Although Ecuador, Brazil, Mexico and Central America are all interested in developing oil palm production, it is Colombia that has a massive lead in this sector. Just to show that the subject is not purely theoretical and how it impacts on the daily life of human beings, the following text, extracted from personal notes on travel in Colombia in July and August 2007, documents the human dimension of the 'great project' of green energy. It brings to mind John Steinbeck's *Grapes of Wrath*.

PERSONAL TESTIMONY FROM COLOMBIA

The Intercongregational Justice and Peace Commission is one of the organizations in Colombia that are constantly being threatened because they are concerned with human rights violations among the peasant populations who have been uprooted from their land, particularly to make way for the extension of agrofuels.

One evening, with a few of their members, I visit an indigenous community, north of Bogotá, halfway up a mountainside to spend a night of prayer. Gathered around this circular sacred place, lit by a fire in the centre, we listen to an old man who described how agricultural companies had expelled them from their lands, during which there were massacres. We pray for the dead. There are long silences. Members of the community join the group. They salute us by touching their foreheads (a welcome from their thoughts) and exchange a few leaves of coca. One by one, they speak, because 'the word is the soul'. The old man who is presiding invites me to speak first, because I too am an old man. It is a spiritual sharing, reflecting a respect for Mother Earth and the importance of human life. There is an acknowledgement also for the local people who had fraternally welcomed them on their own land. A contrast: the first name of the old man is Victor Hugo and before the ceremony he asks everyone to switch off their cell phones! I do not stay the whole night because we had a full day's work ahead of us.

An international seminar on agrofuels is being held the following day and I had been asked to give the opening speech. Among the participants are not only Lain Americans and Europeans but also Asians. Experiences are exchanged and the refrains are similar: the social and ecological effects of the production of green energy have not been taken into account. The next day I spend the day with an international delegation near the frontier of Venezuela at Arauca, by a tributary of the Orinoco river, to hear the testimonies of people, mostly peasants, who had been forcibly displaced from their land. This is a hearing to prepare a session of the International Opinion Tribunal which took place a few months later, in November 2007, in Bogotá. For half a day we hear one account more dramatic than the next, with many of the witnesses having to speak behind a door so as not to be recognized. There are expulsions by the oil companies or agrobusiness, massacres by paramilitaries and the army, bombardment of a village by a helicopter belonging to an American oil company. We interrogate the speakers. A German parliamentarian, overcome with emotion, is incapable of posing a single question. While the delegation is waiting for the plane to take us back, two policemen from the DAS (Security Administration Department) come to take away our passports and accuse us of illegal activities, threatening us with expulsion. We reply calmly that we had informed the Bogotá authorities beforehand. A telephone call confirms this.

We then visit the northern part of Chocó, a region at the frontier with Panama and near the Atlantic coast. Our first plane stop is at Medellín, which has become a veritable metropolis. It is possible to understand the general form of its social structure by over-flying the city at a low altitude and landing at the airport which is at the very centre of the city. There has been a huge development of the districts of the richest, with many tall buildings, while poor districts stretch away as far as the eye can see. It is not, however, as large as Bogotá, a city of 8 million inhabitants, the north of which displays its opulence, while the south is made up of the districts in which hundreds of thousands of people, mostly internal refugees, are crowded together. From Medellín we fly to a small town, Atrato, situated not far from the Panamanian border. There are a number of us undertaking this part of the journey, particularly members of the Justice and Peace Commission, including a Sacred Heart nun and the Ethical Commission of South Atrato, of which I am a member, a Spanish jurist and two young Americans, one of whom had just come out of prison, having demonstrated in front of the School of the Americas in Georgia (a place with a sinister reputation, where Latin American military are trained).

Before landing, we fly over banana plantations. We then cross them in a car before reaching the little local town. From the restaurant I observe the details of everyday life and they remind me of passages of Gabriel García Marquez's *A Hundred Years of Solitude*. From there we take two cars to go to another small town in the interior, where we are received by a convent of nuns, a Colombian congregation founded by Mother Laura, for work with indigenous populations. They are known as the Lauritas and are to be found in several Latin American countries doing social work with indigenous people. This congregation, which is very committed, also extends its activities to include the poorest of the population. We spend part of the evening with the nuns, hearing about their work in this region where the large landowners own thousands of hectares and where today the plantations of African palm continue to be extended. They describe the huge number of peasants who arrive, chased from their land into different towns, especially in the one where we are now. It is called a town but few of the streets have been asphalted and the public services are rudimentary. One part has been hastily built with houses made from wooden planks and zinc roofs by displaced rural people from the interior.

The nuns explain how these peasants have lost their possessions. Having nothing left, they are obliged to manage as best they can. No financial compensation, no social security, great difficulties in sending their children to school, no jobs for the young. The situations are often dramatic and the team of seven nuns cannot deal with them all. One of the nuns says: 'Among the internally displaced people, there are many who are of African descent and also some indigenous people. Often terror has been used to make them leave their land. The paramilitaries, informal armed groups but intimately linked with the army, threaten and assassinate people, just to terrify them.' Another adds: 'The paramilitaries are a veritable state apparatus, because their liaison with the military, political and economic powers is obvious. The present government pretends to demobilize the paramilitaries, but in actual fact they are as present as they were before and as armed as always. A certain number of them who have been released into civilian life go practically unpunished and they occupy very important positions in the political field, even in Parliament.'

The Mother Superior of the little community adds that these political practices are linked to the continual extension of the large landholdings. It was not so long ago that a good proportion of the land was wooded. The rest was cultivated by small peasants or by indigenous communities. Then first came the extensive livestock ranches, afterwards banana cultivation and now it is palm oil. At the beginning it was the military who, using coercion, helped the landowners to get more and more land from the peasants. Then came the paramilitaries to do the dirty work that the military could not do.

We ask them whether the guerrillas are active in the region. They say that they were very active during the 1990s, but that they had now taken refuge in the mountains. At the start they were peasants who had been robbed of their land, organized in armed resistance. Over the last 40 years this resistance, particularly the FARC (Revolutionary Armed Forces of Colombia), has become a military organization which, in order to keep itself in arms, has imposed taxes on the drug trade and gone in for kidnapping people. In this region, there is no particular sympathy for them among the population, displaced though it is, but nor are they considered as enemies because in the regions that they control the peasantry is better off.

Drug trafficking is present everywhere: the paramilitaries live off it and the military are often involved too. Members of the Intercongregational Justice and Peace Commission say that drug trafficking has penetrated throughout Colombian society. A large part of the urban buildings of Bogotá and Medellín are constructed with profits from this source. The government, helped by the Americans, destroys the coca crops through fumigation, above all in the mountainous zones far from the towns, but it is not very efficient and indeed has disastrous ecological effects. The attacks are against the small peasants who, often uprooted from their traditional cultivation, have no other means of existence. But the big drug traffickers succeed in making a place for themselves in society.

We pass that night in premises belonging to the parish, with foam-rubber mattresses on the ground. As the oldest of the party, I was given a bed of planks. After a short sleep, we continue on our way. First along the highway which leads to Medellín and which crosses extensive cattle ranches, spread over thousands of hectares. The animals are scattered and relative few in number. We pass 16 military checkpoints during our two-day journey, which indicates the militarization of the region. At two of the checkpoints we had to identify ourselves. Our jeeps then take to the country roads alongside which we see small houses made of stubble and roofs of corrugated iron. The inhabitants are the families of the paramilitaries who come from other regions.

After travelling for 75 kilometres we reach the zone of the African palm. Now we are travelling along lanes that are completely surrounded by palm trees. They are not very high and their fruit grows near the bottom of the trees. All along the lanes we see the harvested fruit collected together. Lorries come to take them to a nearby refinery. One hectare of palm trees gives 5,000 litres of agrodiesel, so it is a very profitable crop as it does not require a lot of labour. The workers are brought to the spot each day. The fertilizers and pesticides that are used on the plantations are chemically very destructive. There is not a single bird in sight. The peasants tell us that the streams and rivers no longer have any fish. One of them shows us the burn marks on his skin that he received when he had bathed in a river. The chemical products are spread by planes and respect nothing, neither the soil, nor the water, nor the rare spaces where there is still some habitat.

Finally we reach a place where there is a big board on which the words 'Humanitarian Zone' are written by hand. After a decade of bloody struggles led by the 17th Brigade of the army and paramilitaries acting for the oil palm companies, and then, from 2001, successive forcible displacements, a group of peasants joined together to cultivate a few hectares of land on the edge of the palm plantations. They had been expropriated from their ancestral lands, some of them dating back more than 120 years. They created what they called the 'Biodiversity Humanitarian Zone'. In doing so they have been accompanied by members of the International Peace Brigade who are protecting them. An international ethical commission had been set up to alert the international bodies in case of serious violations of their rights. The authorities in Bogotá had been warned of our mission by the Intercongregational Commission of Justice and Peace.

It should be mentioned that a small peasant in these regions was someone who possessed between 50 and 100 hectares. The crops were diversified, cattle raising was relatively extensive and the woods plentiful, all of which enabled these peasants to have a relatively normal life, even if working conditions were hard. As has already been said, Chocó was the region that had the greatest biodiversity in the country. A few kilometres away there had been a village, with a primary school, a health centre, an aqueduct bringing water from the mountain, a series of churches of different Christian denominations. Now there is almost nothing left of the village: the school, the health centre, the aqueduct have been destroyed to make way for the oil palm cultivation. Each increase in the plantations brings new massacres with it. In December 2005 at Pueblo Nuevo and again in October 2006 at Brisas, there was a new paramilitary group, called the Black Eagles (Las Aguilas Negras) who conducted the operation, linked with the army and the police.

We get down from the jeeps and go towards a small group of dwellings that have just been built. The peasants have been installed on this land only a few months ago and already they are the object of a legal action which refers to them as 'invaders'. In fact, all of them used to have land from which they had been expelled. Many oil palm companies are active in the region, stimulated by the boom in agroenergy. Here, it is Urapalma, a limited company.

As the peasants did not want to give up their land, threats were quick to follow. They said 'If you don't want to sell your land, we shall buy it from your widows.' Unfortunately, that is what happened. In the community that we visited, 113 people had been assassinated, first by the army and then by the paramilitaries. The same thing happened in many other places. I will not describe the way in which they were killed: it is beyond my powers of description. Recently, one of them, a black, who was to go to an international meeting in Chicago to denounce the injustices perpetrated in Colombia, was assassinated a few days before his departure. His body was found in the river by one of the Sacred Heart nuns present with us. It was a warning to the others.

There are members of the International Peace Brigade here: young Italians, Spaniards, Americans, Canadians and French, who take it in turns to live and work with the community, running all the risks of protecting them. This is the meaning of the 'Biodiversity Humanitarian Zone' – in other words, it is an area of five hectares of land close to the palm plantations that has been taken over by one of the peasants and which is symbolically protected. Our visit, which is made in the name of the Ethical Commission, also aims at preventing the continuation of the silent and unpublicized atrocities. The government, concerned about its international reputation, fears that these events may be exposed. Some international legal bodies have indeed been alerted.

In the afternoon we walk for about 2 kilometres to the cemetery on the edge of the plantations. It has been completely destroyed by bulldozers and all the tombs desecrated. A small piece of land outside the plantation has been replanted by the peasants, with little wooden crosses painted white. As we arrive along the path, which is full of water and mud, a viper appears and is killed by one of the peasants. Silently, we gather round the place. Someone explains the history of the cemetery. 'We had a village here. There is practically nothing left now. The machines completely destroyed the cemetery. We don't know any more who among our dear ones are here. We have put a few crosses symbolically but without knowing who is below.' Emotions run high.

They ask me to bless the place and to pray for their dead. We recite 'Our Father' together and remain for some time to pray silently, not only for the parents and the grandparents who are buried there, but also for those who have been massacred. Far away we can hear chainsaws at work in the forest.

All this profoundly upsets me. It is difficult not to feel rage when you see such things. Capitalism has no respect for anything. One must make money. Everything has to be transformed into saleable merchandise. That's the supreme value. Human beings do not count any more, even those who are resting in peace beneath the earth in this country cemetery. We return on our tracks through the plantation: the palm trees of death.

That evening we dine together, on bean soup. There is a series of witnesses, songs, a sharing of experiences. Some people from the region of Cacarica have come to join us, having spent a whole day travelling, coming down the river Curvarado, which flows about 20 kilometres away. There are young blacks and peasants from close by: men, women and children and even they have walked two, three, four hours to get here to spend a few hours together.

The evening starts with the witnesses. They recount the expulsions, the massacres. One after the other, peasants tell their story: the threats of the army, the killings by the paramilitaries. One man speaks of his three-year-old child, murdered before his eyes. A young man describes how his parents had been killed by the paramilitaries. Others bear witness in quiet, sad voices. It is extremely moving. We are all very quiet. An old woman of African origin speaks: 'I am a grandmother and I have 29 grandchildren. I was chased away from my land. My grandchildren could not go to school. We have no medical services, not even a little health centre. We are peasants. We want to work on the land. But I so much want my grandchildren to study, to develop their lives. What have we done to suffer all this? We want to live in peace, to cultivate our land. There was life here and now there is only death. Yet, we haven't lost all hope. We believe that the Lord has not forgotten us. We continue to struggle. We will not be discouraged by threats and violence. We want to live in peace.'

The young people from Cacarica express their feelings by singing. It is rap and the author of the songs is among them. He explains that rap originated from the black Americans and that they were songs of protest. They sing several times during the evening, which is spent under a sheet-metal roof, with a beaten earth floor. Their songs are impressive, with their staccato rhythm, telling the story of their community. They too have seen their families expelled from their land and they demand justice. They accuse the big landowners and the agrobusiness companies and they denounce the paramilitaries who have massacred many of their relatives. They accuse the army, the government and in particular President Uribe, himself the owner of much land as well as mines, and who has arranged for the paramilitaries to go unpunished. Some of the songs are very tough, but they end with the desire to struggle and not give way to despair.

When the witnesses finish, they ask me for a prayer. Everyone falls silent. They remember the killings and the victims. The resurrection of Jesus is indeed the symbol of the victory of life over death. God is present. He is the One who is for life. Jesus, too, was persecuted and finally executed because He was opposed to the domination and exploitation of the powerful over the poor. But He rose from the dead and this is the source of hope.

A young black woman starts singing the praises of the black virgin. It is a long chant, sung sweetly, and the whole assembly, moved, listen attentively. 'The virgin is the one who loved, who suffered. She also loved the black people. This is why we call her the black virgin. It is she who gives us hope. It is she who thinks of us, like a mother. The black virgin … the black virgin.' A long silence follows, everyone lost in their thoughts.

But the *joie de vivre* soon returns. All are invited to sing a refrain, to tell a tale. Life is winning over death, joy over sadness. Everyone from the different communities tries to participate in the celebration. Some sing awkwardly, very much out of tune. We all start laughing. The children have prepared a song, which never finishes – but they too have participated. As there are a few foreigners present, they ask the two American women to sing. It's quite a catastrophe. The Italians join in, with greater panache. They also ask me to make a contribution. Not being very gifted in this field, I decide to teach them the French song 'Frère Jacques'. It starts off quite well but what a surprise to see that everyone can sing in French! Inevitably it all ends in a pandemonium. But everybody laughs heartily and it really is a fête. It all lasts until quite late and, like this morning, we have to get up early the next day. Finally everyone returns to their hovels or little tents and the only light is that of the moon, just visible through the clouds. I sleep on a few planks, luckily with a mosquito net because those little creatures had joined us in the party! Beside me, a young Catalan is snoring loudly – well, each man for himself! It isn't easy to sleep, but finally fatigue sets in. At 4.30 in the morning I am suddenly awakened by the crowing of a cock who has seen the first rays of the sun. He is on the other side of me, separated by a sheet of plastic. We all get up. It had rained during the night. The earth floors, even inside the houses, have become mud. Not easy to get dressed. Happily, an enormous bowl of coffee has been prepared. It helps us become fully awake.

At six o'clock we must all get ready for an operation to destroy palm trees. We follow the track that leads to the road. At least a hundred people are present: peasants, members of the brigade, young and old. Everyone has a machete. Before starting, one of the peasants, who has been dispossessed of his land, speaks. He is a member of a Christian denomination which is called the Quadrangular Church (the four corners of the earth). He asks us all to engage in private prayer. I am beside him. He then continues by reciting 'Our Father' and I join in. Then, with his eyes lowered reverentially, he asks for the blessing of God on all those who want justice. 'May God give us the strength to continue to struggle, to re-establish justice, to fight for our family, for life, for fraternity between all people. And now, with our machetes we are going to destroy the work of death.' The group moves into the plantations. Each one chooses a tree and the machetes get to work to hack down the oil palms.

Some of us have to return to Bogotá, so our participation can only be symbolic. We have to get back to the capital. I say goodbye with emotion to those with whom I had shared those very intensive hours.

But there is a problem. During the night a fierce tornado had felled a number of trees across the only road that leads out of the 'Humanitarian Zone'. There is no way to get by with our vehicles. We have to go on foot and, with some people accompanying us, we start walking. Unfortunately, I had hurt my right leg two days previously getting into a vehicle and it was still not in very good shape for walking. But I have an umbrella which serves as a stick and off we go, along paths full of holes and mud. We walk monotonous kilometre after kilometre between the rows of palm trees, when we meet a lorry carrying workers: it had remained within the perimeter of the roads that had been blocked by the fallen trees. After nearly 10 kilometres, a motorcycle arrives. We ask for a hitchhike. The rider is a local peasant. Again, I receive the privilege of age and am told to get up on the motorcycle and so we continue the other 10 kilometres before reaching the river. I still wonder how we managed to avoid having several baths in the ditches along the route! I think of the film *The Motorcycle Diaries* and almost fancy myself as Che!

On reaching the river, the motorcycle owner uses his cell phone to call the little town on the other side of the river. He calls on two other friends with motorcycles who come to take the others who had continued on foot, while the sun begins to beat down intensely. Finally we are all together again and we take a pirogue (a dugout canoe) to cross the river and retrieve our jeep on the other side. Another military block. More than 70 kilometres along impossible routes. Lorries are getting stuck in the mud – everything seems to conspire to make us lose our plane – because the journey has taken us much longer than planned. Finally we get to the main road. But the jeep cannot go faster than 40 kilometres an hour because each ditch in the road shakes it up. We change vehicles in the town where we had spent the first night. We then drive fast towards the airport and happily the plane is an hour late, otherwise we would have missed it.

The return is like the outward journey: a stop and change of plane at Medellín and finally landing in Bogotá. As we travel I cannot help thinking of everything we had experienced during the last two days. The recent events keep passing through my mind's eye. How can such situations be accepted? How is it possible that the church hierarchy is not there to defend the cause of justice? How can a society be built on such grounds? As we passed Medellín, I think of the Upalma company, whose headquarters are in that city. Who are the shareholders? Probably, excellent persons, good fathers of families, good Christians, who sit round a green baize table and take economic decisions according to the logic of profit, without posing other questions. This system must be denounced. The shareholders must be identified. What are the banks that finance them and what are their international connections? They are responsible for killings, the destitution of thousands of people whose human talents are prevented from developing and whose children are blocked from being, one day, able to contribute to the well-being of humanity. They represent material interests against human values.

It could be said that such concerns would mean halting progress, that for a higher good there needs to be sacrifices. But what progress and what sacrifices? Continuing an energy model that chokes our towns and made it possible that, on 18 August 2007, there was a traffic jam 580 kilometres long in France? This, at the price of irreparable damage to biodiversity, to water supplies, to the soil, to the climate, to the detriment of peasant agriculture and for the profits of agrobusiness, dominated by a few huge corporations. More serious still, it is at the price of human, social and cultural sacrifice that affects millions of people. Another model is possible: respect for biodiversity, for human rights and for the climate. But that requires political will.

The day after our return, the police and the army descended on the 'Humanitarian Zone'. Ten hectares of palm trees had been destroyed (out of 25,000, most of which had required the destruction of original forest, thousands of years old). The peasant palm cutters were taken into custody, accused of 'destruction of the environment'. That beats all! The international presence is preventing massacres – for the moment. During the seminar on agrofuels in Bogotá, there is a discussion with the Deputy Minister of Agriculture and a representative of the National Federation of Palm Planters. The latter declares that Urapalma is not a member of the federation and he cannot accept any responsibility for it. He added that the other plantations act with a real entrepreneurial spirit, respecting social responsibility and with a code of conduct. As for the property rights of peasants and of the indigenous and black communities it is, according to him, a complex question because many of the property certificates are false. To verify them all takes up much time and the Colombian state, which has subsidized the plantations, has to recover its expenses and cannot afford the luxury of waiting for years and years. True *langue de bois* (literally, 'wooden tongue', typical bureaucratic language), when it is a question of peasants who have been dispossessed and who are defenceless.

It is indeed a strange discourse, when one knows that between 2001 and 2005, 263,000 peasant families have been expropriated from 2.6 million hectares, either by agrobusiness companies or by the paramilitaries themselves, and that rural poverty increased from 66 per cent to 69 per cent, between 2003 and 2004 alone!

As for the Deputy Minister, he puts forward scientific arguments and declares that Colombia is a model for its respect for biodiversity in the field of palm plantations. To say otherwise would be an insult to the country. Listening to these two speeches, we seem to be hearing something from another planet. What is the logic driving these discourses and their practices? That of progress, represented by the monocultures aimed at meeting the consumption of the richest people in the world? They will soon produce the 'green energy' that everyone talks about but that in reality seems to cause more ecological and social destruction, than the advantages it is said to bring. It is also the logic of profit, because the plantations represent much more added value than peasant agriculture and thus contribute to the accumulation of capital. People speak about real socialism, why don't they speak about real capitalism?

This is of course an anecdotal testimony. But those voices join with those of many others, as thousands of hectares of palm oil trees cover the plains, after having destroyed forests and chased out the peasants of Colombia, Ecuador, Costa Rica, Honduras, Chiapas (Mexico) – and, crossing the ocean, of Cameroon, Nigeria, Congo, arriving at Indonesia and Malaysia and finally Papua New Guinea; the sad soya monocultures that have eliminated biodiversity and cleared human beings out of the countryside in Paraguay, Argentina and Brazil: the 'bitter sugar' which is converted into smelly fuel, the product of slave labour. All this ends by creating a growing buzz which will gradually become a deafening roar, mixing together the cries of the land and the cries of the oppressed which soon the arguments of economic rationality will no longer be able to suppress. An apocalyptic discourse? That is precisely what we would like to avoid. Is it the excessive language of an observer who is so involved that he loses his balance? An anti-capitalist prejudice, forgetting that for all progress a price has to be paid? Let readers judge for themselves. And, because we must, we return to a more analytical language.

FROM *JATROPHA CURCAS*

ORIGINS AND CHARACTERISTICS

The *Jatropha curcas* bush, also called pourghere, is at present considered the most promising among the perennial oleaginous plants suitable for plantations in savannah land and marginal ecosystems in developing countries.[8] The idea of using it for agrofuel is not recent. During the Second World War, in 1942, the French colonizers experimented with it, foreseeing the scarcity of petrol. The tests were inconclusive and the project abandoned. But at the beginning of the 1990s experiments were started up again, with the installation of an engine run by agrofuel to operate a flour mill and an electricity generator. Since then, various studies have confirmed the technical and economic feasibility of this plant, and its environmental advantages.

Jatropha curcas originated in Latin America, where it was known by the Mayas for its medicinal properties. It crossed over to Africa via the Cape Verde islands and Guinea Bissau before being introduced by the Portuguese to other parts of Africa, as well as Asia. From the seventeenth century, Thai peasants were being encouraged to cultivate it for the oil in its grains, which make a soap prized for its special foamy qualities, as well as for lighting oil lamps. In the Democratic Republic of the Congo it was not an indigenous plant but can be found in small quantities everywhere in the country and it mainly serves as hedges separating plots of land and making pens for livestock. The plant has now spread to all the tropical regions and it is to be found in Mali, Burkina Faso, Senegal, Niger, Madagascar and Egypt, as well India, China, Vietnam and Thailand.

A major reason for the popularity of jatropha, among the energy sources from biomass, is that its cultivation and exploitation have none of the disadvantages of colza, sunflower, soya or oil palm. In India, as has been reported by the Society for Rural Initiatives for Promotion of Herbals (SRIPHL) in Rajasthan, the government has chosen the *Jatropha curcas* for several reasons, particularly for the low cost of its grains, its high yield in oil, its adaptation to various climatic conditions and its grain harvest, which takes place in the dry season.

In contrast to the oil palm, a somewhat 'bourgeois' plant that involves massive deforestation, jatropha seems 'proletarian' because anyone can cultivate it. Its costs of exploitation in plantations are far fewer than those of the oil palm. Another ecological advantage is that it is an effective agent against soil erosion.

At the international level, the advantages of *Jatropha curcas* have aroused the interest of companies from the UK and the USA, and other transnationals. But in spite of its oleaginous properties and its other advantages, it has not yet been exploited industrially. However, the projects are multiplying. Research is under way in Africa and India, often by Western companies in collaboration with universities and local governments, which have launched projects that are gradually attracting careers for their own nationals.[9] Various experiments in intensive cultivation have been made by the oil companies and by some governments.

CONTEMPORARY PRODUCTION PROGRAMMES IN ASIA AND AFRICA

India has launched a vast programme for the plantation and selection of cultivars with the highest yields and hence its interest in *Jatropha curcas*. It plans to plant some 40 million hectares between now and 2012. Plantations are also being developed in other Asian countries (such as Indonesia, China, Vietnam, the Philippines and Thailand). In Indonesia the price for its seeds has rocketed: a tonne of this much-sought-after grain costs up to 1,000 dollars in 2008. In December of that year the government of Myanmar issued a decree aiming to create plantations of *Jatropha curcas* for a total of over 200,000 hectares in each of its 14 provinces to replace, within three years, part of its consumption of diesel oil. It is estimated that the production of agrodiesel would then be 227,300 million litres per province. In Surinam, the American company Tropilab has announced its intention to plant 50,000 hectares of the plant and then to build a refinery to transform the crude oil into agrodiesel in situ.

A conference on jatropha was held in June 2006 in India. Participants advocated measures that would attain a high level of energy independence: at present India imports 73 per cent of the 125 million tonnes of crude oil that it consumes each year, and its own reserves can only last another 20 years at the present rate of consumption. In a first stage, out of its 296 million hectares of uncultivated land, it is planned to dedicate about 40 hectares to this culture. According to a report of the Indira Gandhi Agricultural University, 'the culture of *Jatropha curcas* can make a significant contribution to the production of agrofuels and to the sustainable development of the country ... as beneficial to the cultivators as to the industrialists ... Jatropha is unique among the renewable sources of energy, as it is simple to cultivate and requires a relatively low investment.'[10]

Since the Kyoto Protocol entered into force, it has added a juridical argument concerning the quotas of CO_2 emissions. India and certain African countries became aware of the potentialities of *Jatropha curcas* and hurriedly launched large programmes, counting on the possibility of receiving capital support from the Certified Emission Reductions (CERs).[11] In fact, such plantations have great advantages in that they enable hitherto uncultivated land to be reforested and, in certain conditions, food crops can be cultivated between the trees which, by their shade and humus, fertilize the soil. Vegetable crops like cucumbers, tomatoes, lettuce and pumpkins can be grown. This makes it possible to compensate for the expenses incurred during the years when the costs of the plantation are greater than the income.

Other conditions are favourable for the development of *Jatropha curcas*, according to its promoters, particularly as there is a certain tradition in cultivating cash crops (palm oil, coffee, cacao trees, sugar cane, etc.). Massive public and private investments in jatropha (commercial and village plantations) could help resolve the problems of poverty and unemployment, according to the same information source. The development of the sector would both meet needs and the demand for diesel in the country and therefore relaunch some industrial sectors to ensure consumption from transport. The country can thus respect its obligations in the Kyoto Protocol. If only 3 per cent of the surface of the Congo were planted in jatropha, some 7,350,000 square kilometres, with a minimum production of 1.5 tonnes of oil per hectare, the country would produce 11,025,000 tonnes of crude oil, which would represent an annual income of about 5.5 billion dollars.[12] These are arguments similar to those put forward for ethanol from sugar cane, but none of them take ecological and social 'externalities' into consideration.

Other African countries, particularly South Africa, intend to invest in agrofuels in the hope of obtaining a share of the market in the EU. In Tanzania it is estimated that almost half the national territory could lend itself to this type of production and the government is looking for investments from European producers such as Sun Biofuels in the UK. In Mozambique, almost 33 million hectares – about 40 per cent of the country's territory – have been identified, especially for supplying the European market in agrofuels.

In Mali, Aboubacar Samaké, President of the National Programme for the Energy Exploitation of the Pourghere Plant (PNVEP) has much hope for this plant. According to him, the cost of production of a litre of oil from *Jatropha curcas* could be estimated at between 170 FCFA and 250 FCFA as against 475 FCFA, about double, for diesel (1 US dollar = 482.92 FCFA). This programme could eventually be a source of economies, particularly by reducing the importation of oil. Besides supplying vehicles, it could also provide electricity for the rural areas. The immediate plan envisages the electrification of five villages in five years. Also planned is an increase in the national production of pourghere grains, thanks to the work done on the perimeters by rural populations. On the 45th anniversary of the country's independence, President Amadou Toumani Touré declared: 'The promotion and exploitation on a larger scale of agricultural fuel derived from bagani, the pourghere plant, should be taken into account among alternative strategies.' It is even being considered to create a development company to improve its yield in grains, its cultivation cycle and the performance of the oil produced as fuel. An interesting example of the local utilization of the agrofuel is the testimony of Batou Bagayoko, chief of Kéléya in Mali, who does not hide his satisfaction in seeing the roads of his village being lit at night, electricity being very rare in rural Mali. Situated some 100 kilometres to the south of the capital Bamako, Kéléya is the first village to benefit from electricity provided by a generator using oil from pourghere (jatropha).[13]

The Belwet Association in Burkina Faso works together with the Nature Tech Afrique (a company specializing in renewable energies) and the German agrodiesel company Deutsche Bio Diesel (DBD), particularly on how to transform jatropha grain oil cakes into livestock feed. According to Larlé Naaba Tigré, President of Belwet, this plant makes it possible to restore the soil and develop a rural market for the oil, promoting its use as a domestic fuel for cooking food so as diminish the use of wood for this purpose, as well as developing the manufacture of soap and thus create extra income for the peasants.

In Madagascar, the sale of jatropha oil constitutes an extra revenue estimated at 1 million ariary (472 US dollars) per peasant and per year (60 per cent of the Malagasy population live on less than 1 dollar a day). 'On top of its energy virtues, the jatropha could replace charcoal, be used as fertilizer and make candles and soap', says Sally Ross, Director of DL Oils Madagascar, a company based in London which is closely following that country and all its flora resources. Replacing diesel by the green fuel should be all the easier because the technology can be used 100 per cent by diesel engines and the price per litre of this oil would be equivalent to that of ordinary diesel oil. Plantations were under way at the end of 2008 in three regions in Madagascar on a total surface of 1.631 hectares with the participation of 1,500 peasants.

As for mega-projects, it is monoculture that is proving to be the main way of producing *Jatropha curcas*. Hence the peasantry will be transformed into a rural proletariat and cultivation will extend well beyond the arid areas. As has been said, the plant's yield increases considerably on better land. Another way of ensuring the future of this sector of green fuel would involve the peasants themselves in the process. (This is also happening in certain regions in Papua New Guinea and Colombia with the oil palm.) Sally Ross explains: 'the company gives the young plants to the farmers, who harvest the grains and crush them, extracting the oil which is bought by the company which refines and transforms it into green fuel'. In other words, the peasantry simply carries out the work and is totally dependent on the companies whose interest it is to increase profit margins and thus control the costs.

There is, however, a different development model, based on peasant production and which aims at satisfying local needs, a solution that also has the merit of respecting biodiversity. It is therefore totally different from the ones that we have described and which are being rapidly set up without consideration for ecological and social costs.

The fact is, though, that *Jatropha curcas*, which could be developed on behalf of the local population and under public control of the cultivation on land that is unusable for growing food crops, has fallen into the hands of agricultural companies, often transnationals. Not only is biodiversity not respected and arable land transformed into monocultures but, according to an Oxfam report on Tanzania, groups who are socially vulnerable are already being expelled from their land. Once this logic is introduced the cultivation of this plant reproduces all the ecological and social mechanisms already described for the cultivation of sugar cane and the oil palm.

It should be added that in-depth scientific studies do not yet provide guarantees of success and that in some cases unexpected consequences, such as the proliferation of certain noxious insects, could cancel out many of the advantages. During the international conference on the issues and prospects of agrofuels that was held in Ouagadougou, 27–29 June 2007, at the round table on jatropha questions were posed that could not be answered. 'Is this a plant of the future for large agro-industrial exploitations or for family cultivation? How much water does it need ? How does it compete with other plants? How does its production vary? What varieties should be used? What is the rate of success of plantations? How should the cultivation be carried out? What resistance does it have to insects, such as ants? How does it fare in industrial plantations?'[14] What could be its impact on the ecosystem?

There are so many questions that still remain without answers. Analysis of its toxicity is going to be indispensable because the Institute of Development Research at Noumea in New Caledonia has reported cases of intoxication from jatropha grains. The main risks for children are dehydration, cardiovascular collapse and depression of the central nervous system, even though no deaths have been formally registered. Dr Kurt Hostettmann has stated that the plant has esters of phobol with toxic properties, particularly harmful for eyes and skins. This was confirmed in a laboratory by researchers at the University of Graz in Austria, in 1999. Researchers at the University of Wageningen in the Netherlands also said that up until now they did not have sufficient data about the real potential of jatropha in less than optimal conditions. What it means is that there are other considerations to be taken into account rather than simple profit, such as needs, on one hand, and precaution, on the other.

PLANTS SIMILAR TO *JATROPHA CURCAS*

Jatropha is not the only plant in the countries of the South that can produce agrodiesel and which have certain advantages. Among them are the *Moringa oleifera* (drumstick tree) and the *Millettia pinnata* (karanj).

The *Moringa oleifera*, often called just moringa, is a tree belonging to the *Moringaceae* family that can grow up to 10 metres. It comes originally from north India and has now spread over almost all tropical regions. It resists drought well and grows rapidly. In India, the moringa is cultivated for its fruit, which is cooked and exported, fresh or conserved. In the Sahel its leaves are eaten as vegetables and those of *Moringa stenopetala* are the staple food of the Konso people in Ethiopia. Nutritional analysis have proved that the leaves of *Moringa oleifera* are richer in vitamins, minerals and proteins than most vegetables. They constitute a complete diet, as they contain twice as many proteins and calcium as milk, as much potassium as the banana, as much vitamin A as carrots, as much iron as beef and lentils and twice as much vitamin C as an orange. Many programmes utilize the leaves of moringa against malnutrition and its associated diseases (blindness, etc.). Moreover, its grains contain a cationic polyelectrolyte which has been efficient in treating water (elimination of muddiness), instead of sulphate of aluminium or other flocculating agents. There is a double advantage in using these grains. First, the plant is an easily accessible local product that can substitute the flocculating agents that are imported, which is an important saving in foreign currency for the countries of the South. Also, contrary to sulphate of aluminium, it is completely biodegradable and it is possible to extract from its grains a useful edible oil, especially in Africa where many countries lack edible oils.

As for the *Millettia pinnata*, it is a tree belonging to the *Fabaceae* family. It grows rapidly, fixes azote, is very resistant to drought and flourishes under the hot sun, in difficult, even saline soils, and it produces oil. It is more known under the name of *Pongamia pinnata* but recent genetic studies name it *Millettia pinnata*.

Programmes to plant this tree in Uganda and Cameroon (in the region of Kumbo) are being made on the initiative of the Himalayan Institute of Yoga Science and Philosophy. The tree also has great potential in the fight against desertification, particularly in the Sahelian zone. It is possible to plant 200 trees on one hectare of land and each tree can, starting from the sixth or seventh year, produce from 25 to 49 kilograms of fruit, the oil content of which is from 30 per cent to 35 per cent. One person can harvest 180 kilograms of fruit in one day, working for eight hours. The average yield is 5 tonnes per hectare per year as from the tenth year. Contrary to the *Jatropha curcas*, for which it is necessary to wait three years before obtaining oil, the *Millettia pinnata* supplies some oil as from the first years. The oil cakes (*tourteaux*) obtained after oil extraction make excellent fertilizer.

There are other perennial plants which are creating interest, including the babassu palm tree (*Orbigniya speciosa*), the coconut palm, the shea or karite tree, the castor oil plant (*Rucinus communics*), the neem tree (*Azadirachata indica*), the argan tree (*Argania sideroxylon*) and the cardoon (*Cynara cardunculus*),[15] a grassy herb-like panic or switch grass which has much promise: it resists drought and needs fewer inputs than maize[16] for producing ethanol and is currently being used in the American Midwest to fight erosion through the Conservation Reserve Program.

As can be seen, there are many sources of oil that can be used as fuel. Everything depends on the conditions of exploitation.

7
The Collateral Effects of Agrofuels

Today agrofuels are increasingly being seen as a very partial solution to the exhaustion of the world reserves of fossil fuels and the planet's climate crisis. As has been shown, their exploitation on a world scale has perverse effects, both ecologically and socially, which reduce their effectiveness and the advantages of applying them.

THE ECOLOGICAL EFFECTS OF AGROFUELS

The destruction of primary forests and in general the introduction of monocultures have serious ecological consequences on rainfall and ground water systems, soils and the environment in general. The situation has become worse these last years, with the emergence of new agricultural sources for supplying fuel.

On Water

The conversion of primary forests (in the Congo basin, the Amazon and the forests of South-East Asia) into plantations upsets ecosystems and hence the water cycle, altering the level of rainfall in the regions concerned and even in those further away. The recourse to monoculture leads also to a massive and intensive use of phytosanitary products (pesticides, fungicides, etc.) and fertilizers, which for the most part are mineral-based, such as diuron, metasulfuron, glyphosat, cypermethrin. Fertilizers and pesticides used in monoculture (for example for oil palm plantations) are particularly responsible for the contamination of water, both surface and underground, as has been observed in Indonesia and Malaysia.[1] The lowering of the water table is also a consequence that has been reported in many places in Brazil and in Indonesia, because of the monoculture of sugar cane and the oil palm. According to Parizel, the production of one litre of ethanol from maize requires between 1,200 and 3,600 litres of water.[2]

On Soil

The replacing of primary forest by secondary forest (oil palms, eucalyptus) or by other crops (sugar cane, maize) for the production of agrofuel has ruptured in the soil–water equilibrium. These two elements are in symbiosis with each other in nature. In fact, the soil is protected by trees from fierce battering by rain drops. As a consequence, erosion starts to appear in the soils left naked or incompletely covered by monoculture.[3] A forest also helps to stabilize the soil and hence reduces the erosive effects of rainfall and streams. In the US it has been reported that maize monoculture causes more erosion than any other crop. The farmers in the Midwest, having abandoned the rotation of crops for growing soya and maize exclusively, have experienced increased soil erosion. The lack of crop rotation has made cultivation more vulnerable to different diseases and hence necessitates an increasing amount of pesticides. In the US, 41 per cent of herbicides and 17 per cent of insecticides are applied to maize cultivation.

Cultivation on certain kinds of land causes salinization and acidification. A report by Robert Jackson et al., published by the magazine *Science* in 2005, explained that the replacement of forests by eucalyptus in the pampas of Argentina has resulted in soil salinization. The plants dig deep into the water, thus transmitting the mineral salts that have dissolved on the surface. The consequences are even more dramatic in the dry season when the level of water courses close to the cultivated areas is substantially lower. This is also the case in Minas Gerais in Brazil. According to the same author, the result is an imbalance in the mineral nutrients of the soil, leading to the disappearance of calcium, magnesium and potassium, because they have been too much absorbed by the plants. It has also increased the sodium content so that soils become increasingly saline[4] and hence unsuitable for agriculture.

Furthermore, numerous measures carried out on different soils in Africa, Asia and Latin America have shown that the intensive use of pesticides and fertilizers on oil palm and maize plantations, as well as in other monocultures to produce agrofuels, leads inevitably to soil acidification, so that for a long time they remain unsuitable for any other use.

On the Global Environment

The effects on water and soil also have an effect on climate change at the planetary level which is causing massive destruction of the tropical ecosystems.[5] Devoting tropical forest land to other cultivation leads in the medium and long term to climate changes that cannot be ignored at the world level. There is, in fact, an interaction between the three elements: water, forest, climate. It is a delicate symbiosis so that an incautious manipulation of one of the three brings about an imbalance of the system as a whole which can be incommensurable. The interaction of these elements can have large-scale implications: according to a study at Oxford University, it appears that the uncontrolled deforestation of the Congo has reduced the rainfall in the Great Lakes region of the US (by 5–15 per cent), as well as in Ukraine and Russia, to the north of the Black Sea.[6] In its turn, climate change seriously affects the tropical forest where a fall in precipitation has been recorded in recent years.

Another study, carried out in Switzerland, shows that the production and manufacture of agrofuels risk being still more harmful than petrol or diesel from fossil fuels. Hyper-fertilization and acidification of the soil causes loss of species biodiversity.[7] According to the same study, one of the dangers of agrofuels for the environment is at the level of the production of the raw material itself. In tropical Africa, for example, one of the methods used for extending oil palm cultivation is 'slash and burn' agriculture which immediately creates the emission of a large quantity of CO_2 and of soot, which increases air pollution. At Minas Gerais in Brazil, there is a similar phenomenon in the transformation of eucalyptus into charcoal by the steel industry. This leads to a reduction of the soil fauna, which is important for its structure, as well as for fixing atmospheric azote. As a result, the soil is laid bare, which makes it susceptible to erosion. Finally, soil fertility is reduced and even desertification sets in when there has been a longer and more intensive application.

In Indonesia, the conversion of primary forest into oil palm plantations makes a considerable contribution to the release of CO_2 into nature. As we have seen, two types of zone are used for plantations, the dry ones and the marshy ones. In fact, after having been developed in the dryer zones after the forest had been felled, plantations move towards the marshy zones of peat bog. In drying, the latter discharge more carbon oxide into the atmosphere than that absorbed by the trees.

As for the animal world, there has been a huge reduction in the orang-utan population in Indonesia. It is estimated that there had been some 300,000 of them, and now there are only 50,000 left. In 20 years, 80 per cent of their habitat has been taken over by oil palm plantations. The rhinoceros of Sumatra, the tigers, gibbons, tapirs, proboscis monkeys and thousands of other African and Latin American animal species could suffer the same fate.

According to a report published by the Friends of the Earth in September 2008, it is calculated that between 1985 and 2000 the development of oil palm plantations has been responsible for 87 per cent of the deforestation in Malaysia. Even the famous national park Tanjung Puting in Kalimantan has been smashed to pieces by planters. The Delft Hydraulics Institute in the Netherlands considers than each tonne of palm oil is responsible for 33 tonnes of carbon emissions.

Unfortunately, nothing seems to be halting the process and the long-term effects threaten to weigh heavily on climate change because of the loss of these carbons sinks, the tropical forests.[8]

THE SOCIAL EFFECTS OF AGROFUELS

We have also seen that the social effects of agrofuel production are particularly serious. As the situations are very different from one region to another, we shall look at them as they affect both the countries in the North and those in the South.

Agrofuels have generally been welcomed by farmers and political decision-makers in the North, because they generate jobs and are considered an opportunity for family farmers and, above all as a means of reducing their dependence on a barrel of oil that is increasingly expensive and is produced abroad. Some farmers in the North are satisfied, because agrofuels are one of the factors that increased the price of agricultural raw materials for the producer (i.e. of maize in the US), after many years of stagnation in prices. Also it enables them to use arable land that had been set aside, in conformity with the production quotas imposed by the Common Agricultural Policy (CAP) of the EU. But others also see it as creating an increasing dependence on the large companies that control the prices and the market mechanisms. There are therefore the two reactions, the respective strengths of which have not yet been established.

However, it is in the South that the effects are more harmful because it is there where most of the production is concentrated. For, out of the 16 million hectares that Europe needs to supply its agrofuel factories and feed the livestock consumed by its population, only 13 per cent of this land is situated in European countries (according to Friends of the Earth).[9] In fact, the ever-increasing demand for agrofuels at the world level is coming into conflict with a more rational organization of the planet. The EU, with its objectives of 10 per cent of agrofuels to be incorporated into diesel between now and 2010, and 20 per cent of renewable energy in 2020, will need to extend the land to be sown if these objectives are to be met. As it does not have enough land, it will have recourse to the countries of the South, which are currently supplying more than 50 per cent of agrofuels at the world level. This poses the question of supplementary arable surfaces to be allocated to cultivation for agrofuels, while the countries of the South are increasingly confronted by the thorny question of their food security. Furthermore, as we have seen, this situation is responsible for the expulsions and expropriations of many peasants, especially indigenous populations, from the lands of their ancestors. Any resistance to the expulsions or expropriations is met with repression, sometimes killings, particularly by paramilitary forces. This causes huge movements of people towards the great urban centres where peasants will swell the number of the unemployed in the slums, very often living in situations of extreme precariousness.

According to the United Nations Permanent Forum on Indigenous Issues, some 60 million people in the world risk being expelled from their land to make way for the crops necessary to produce agrofuels.[10] Others must remain in the plantations to work in deplorable, sub-human conditions that do not respect the fundamental rights of workers. The women workers particularly are more discriminated against and even paid less than the men.

It is true that the expulsions of peasants started before agrofuels expanded. This was the case, for example, in the 1970s in Paraná in Brazil, when 2.5 million people were displaced for soya cultivation to be used for cooking oil, as well as in Rio Grande do Sul, where 300,000 people had to leave their land for the same reason.

In all the countries of the South, particularly in Latin America and South East Asia, cases are described in reports by the World Rainforest Movement, which is based in Uruguay.[11] Mention has already been made of Western Kalimantan in Indonesia. There the gardens of the Dyaks, which produced wood, honey, medicinal plants and fruit, have been destroyed and the local people now have to cultivate the oil palm trees. Their incomes have fallen and they fluctuate with the prices on the international market. In the same country, in eastern Sumatra, there are 10,800 families who have been forced to emigrate by the PT Citra Mandiri Vidya Nusa, a company belonging to the former Minister of Agriculture. Two years after the beginning of the cultivation of oil palm, the populations displaced by the Mong Rethiby Investment Cambodia Oil Palm company have still not received new land. In the Cameroon, people have been removed their land, without consultation, and reinstalled in new areas with promises of compensation by the companies, who have not kept their word. Customary rights have not been respected and some of the traditional chiefs have been bought or deceived.

These practices are provoking many conflicts. In Cambodia the peasants burnt down 500 oil palms in 2004, making a loss for the company of 70,000 dollars. In Indonesia, in 1998, at Kuala Batu, the peasants set fire to a camp of workers and 49 of them were arrested. Four employees of the Sarawak Oil Palm (in fact, they were members of private security companies) were killed and the Dyaks accused were brought to court. In the same country, the military chased people from their land on behalf of the Tanjung Katung Sejaktera and PT Dasa Anugeran Sejati companies. Similar incidents have occurred in Malaysia, the Philippines, India, Nigeria, Ghana and Papua New Guinea. In Colombia, in the region of Curvarado, massacres have occurred, as has been described in an earlier chapter.

The indigenous populations are among the most vulnerable people. We have already mentioned the Dyaks in Sumatra, but on the same island, in the area of the national park of Bukit Tiga Puluk, indigenous people lost 3,000 hectares of their land, which led to a serious conflict that is still not resolved. In Paraguay the illegal deforestation of the Aroleyo is being carried out on the land of the indigenous people. In Sur Bolivar in Colombia the Afro-descendant communities are being expelled. In Myanmar, the Yan Maing Myint company in 2006 drove away ethnic minorities with the help of the army and the clearing of the land for the plantations was done with forced labour.

Unions are usually forbidden, both in Asia and Latin America, and when they exist they are subdued by repressive measures that prevent them from defending the interests of the workers. In Colombia, many union leaders in these sectors have been assassinated. Neverthless, in September 2007, more than 200,000 sugar cane workers went on strike in the Cauca Valley to obtain more humane working conditions and to protest against the extension of the sugar cane monocultures which were encroaching on rice-producing land and tropical forest. In June 2007, in the state of São Paolo in Brazil, the cane cutters organized a strike to demand a working week of 30 hours and payment by the metre rather than by the tonne.

As can be seen, the social consequences of the extension of agrofuels are very grave. The process forms part of the logic of the exploitation of labour as a factor of production at reduced cost. Like the ecological destruction, the social effects belong to the 'externalities' of economic calculations and the requirements of capital accumulation dominate decision-making.

Finally, it has to be said that, in the countries of the South, cultivation of the various sources of agrofuels (oil palm, eucalyptus, etc.) constitutes revenue on a short- and medium-term basis that is not negligible for the countries that promote such policies in spite of their social consequences. As a result there is a reinforcement of social inequalities and an additional source of corruption. Before drawing lessons from all this, in the next chapter we shall take a look at the socio-economic dimensions of agroenergy.

8
The Socio-economic Dimensions of Agroenergy

THE AGRICULTURAL MODEL UNDERLYING AGROFUELS

As agroenergy is developing today within a capitalist framework, it is perhaps useful to recall the general conditions of agricultural production as a factor in accumulation. The production of food is evidently essential for the survival of humanity. But now there are serious worries about the possibility of feeding all human beings in the medium term. However, the FAO formally declares that the earth can feed 12 billion people. This problem was already brought up in 1963 in a book that I wrote jointly with Michel Cépède, who was the French representative to the FAO at the time, and Linus Grond, the Secretary-General of FERES (the International Federation of Socio-Religious Research Institutes).[1]

Demographic projections have forecast a population of 9–10 billion people by 2050, with the statistics remaining stable from then on. How is it, then, that with some 6 billion people today, more than 1,000 million are suffering from hunger and every 4 seconds a human being dies of hunger?[2] Apart from natural causes, the main reason is of an economic-political nature. This was attested by a report by Fred Magdoff, professor in agronomy at the University of Vermont in the United States: 'Chronic malnutrition and food insecurity are essentially caused by poverty and not by insufficient food production.'[3] It becomes important to pose these questions concerning agrofuels, because they will inevitably compete with food production and because they relate to the dominant logic of agricultural activities.

What is the agrarian economy model promoted by the contemporary economic system? The key issue is precisely that of world food. Because of the hunger problem, the reasoning goes, only increased production can meet the needs. The same line of thinking holds that the small peasant unit is very inefficient from this viewpoint. It therefore becomes necessary to promote an agriculture that is capable of producing enormous quantities. It is also necessary that food habits be changed and become uniform under the influence of 'globalization', as it is not only food that comes from agriculture, but also industrial inputs from the pharmaceutical and cosmetic companies that are causing the soaring demand for agrofuels. How then can this double need be met, that of feeding humanity and producing raw materials and green fuels?

The answer is relatively simple, according to this logic: monoculture must be extended, which makes it possible to diminish the costs by economies of scale and reduce the use of labour thanks to mechanization. Thus it is necessary to concentrate the ownership of land and carry out agrarian counter-reforms. Increasing productivity is needed, hence the spreading of fertilizer to enrich the soils, the application of chemical products to destroy the parasites and the utilization of genetically modified organisms that make plants more resistant to natural and artificial hazards. It is also advantageous to universalize certain breeds of livestock, large, small and medium, because that will facilitate trade. As agricultural labour is abundant, it can also remain cheap. This is the law of the market. The industrial treatment of agricultural and animal products enables the rationalization of production processes and their distribution into the international circuits, just like industrial products.

All this makes it gradually possible to create a world agricultural market, basing prices on the most efficient producers and thus contributing to the rationalization of the agrarian economy. It amounts to a veritable green revolution, on the same level as the industrial revolution. To achieve the Herculean task of being able to feed from 9–10 billion people by the middle of the twenty-first century, only the large companies, capable of transcending national frontiers, are able to meet the challenge. This is essentially the position of the capitalist model. Using these arguments, the discourse becomes moralizing and almost messianic, above all when it is a question of agrofuels. The less polluting nature of these types of fuels compared with energy of fossil origin and hence less destructive of the climate enables them to be qualified as 'biofuels' in the symbolic meaning of the word.

This discourse has a certain logic, but there are considerable grey areas. First, like all capitalist economic reasoning, externalities are not taken into account. As long as profits on investments are not affected by pollution of soil, water and atmosphere, or by the collective cost of uncontrolled urbanization, or by the resistance of peasants who have been evicted and displaced, all these factors are ignored. Differentiated agricultural markets responding to the requirements of food sovereignty and cultural habits, or the collective ownership of cultivable land by peasant and indigenous communities (*ejidos* in Mexico, collective land in Vietnam and China, communal ownership in Sri Lanka) constitute heresies for the capitalist market economy. They should give way to solutions that are more rational and more economically efficient.

But that is precisely the question: efficient in function of what? The ecological balance of the planet or the well-being of peasants (who, let us remember, represent nearly half the world population)? The sustainability of agricultural production, or the accumulation of capital, which is not only a means but an end in itself and limits itself to the short and medium term? This last concern cancels out everything that does not contribute to profit and to its inseparable components: the commodification of all human activities (reduced to their exchange value), obligatory financial profitability, ruthless competition, the entrepreneurial spirit linked exclusively to private ownership of the means of production, the centrality of money that itself has become merchandise.

It is a perversion of what one could call the constants of the economy. It is true that none of these can be neglected: the market is a good regulator of supply and demand when the social relationships between partners is equitable. Profitability, including all the parameters of human well-being, makes it possible to avoid waste. Competition challenges paralysing monopolies and reinforces efficiency in the production process. Finally, the entrepreneurial spirit should be the prerogative of all human beings within a framework of socialized appropriation of the means of production (which does not mean either a simple widening in the number of shareholders or the state necessarily taking over all sectors of the economy). As for the utilization of a universal instrument for the exchange of goods and services such as money, it is a useful means of transaction (without, however, being the only one).

This means that the elements that make up economic activity should be subjected to a criterion that gives them a meaning and attributes them a place within the structure of the system of production and trade. For one model, the well-being of humanity, that is the physical, cultural and spiritual life of all human beings across the world, constitutes this criterion. The use value, rather than the exchange value, is privileged (that is to say, how it contributes to life and its reproduction) and it asserts that certain sectors should not be assimilated to merchandise, that they cannot be measured by the yardstick of purely financial profitability or subject to patents and monetary competition. These sectors include, for example, water, seeds, public services, particularly health and education. It is for each society to define democratically the frontiers of these sectors of the common good, which can of course evolve in time and in space.

In the other model it is the accumulation of capital, considered as the main engine to make the economy function, which forms the basic parameter. The discourse thus identifies the different components of the economy, according to this criterion, as if they could not exist otherwise. Hence the dogma of the market, leading to what is called '*la pensée unique*' (the one and only way of thinking) and the strong conviction of those who claim that there are no alternatives. Worse still, the system makes itself appear natural, refusing to accept that it is a social construction. And yet, as the famous UNDP diagram called the champagne glass, showing the distribution of income in the world (20 per cent of the richest taking 82.4 per cent of the world wealth, and 20 per cent of the poorest sharing 1.6 per cent), it is a minority which, at the world level, monopolizes consumption and the power of decision over the economy, while the 'useless masses' are reduced to bare survival and destitution. But the production of wealth in the world could enable everyone, not only to enjoy their lives, but also to be actors capable of contributing to the well-being of all.

How can that be applied in the field of the organization of the agrarian economy? Samir Amin formulates the issue very well:

Capitalist agriculture, represented by a class of newly rich peasants, not to mention the modernized large estate owners and ranches exploited by agrobusiness transnationals, is getting ready to destroy peasant agriculture. It received the green light of the World Trade Organization at Doha. Production is shared between two sectors whose economic and social nature are quite different. Capitalist agriculture, controlled by the principle of the profitability of capital, which is almost exclusively found in North America, Europe, the Southern Cone of Latin America and Australia, employs hardly more than tens of millions of cultivators who are no longer really 'peasants'. But their productivity, because of motorization ... and the area at their disposal, can produce 10,000 to 20,000 quintals of the equivalent of cereals per worker and per year. The peasant farmers ... who are in turn divided between those who have benefited from the green revolution (fertilizers, pesticides and selected seeds) although with little machinery, whose production amounts to 100 to 500 quintals per worker, and those who have not been through this revolution, whose production is only about 10 quintals per worker. The gap in productivity between the better equipped and the poor peasant agriculture, which used to be 10 to 1 before 1940 is today 2,000 to 1.[4]

This description of the current situation makes one wonder about the mechanisms that have been set up to arrive at this result, such as priority to exportation, inaccessibility of small peasants to credit, importation of food products, massive deforestation, monoculture and concentration of ownership. As an actor in this field, Blairo Maggi, Governor of Mato Grosso in Brazil, is a large-scale producer of soya. He does not beat about the bush, stating that 'the whole economy tends towards concentration. Unit prices fall and we need to make huge amounts in order to survive.'[5] The social consequences are immense. Abourahmane Ndiaye, of the University of Bordeaux, sums it up: 'Twenty million efficient producers ... having substantial production machinery, create 5 billion excluded peasants. The creative dimension of the operation is only a drop compared with the ocean of destruction that it requires.' These phenomena have been admirably analysed in an abundant literature by authors such as Samir Amin (already quoted), Marcel Mazoyer and Jacques Berthelot, which has been largely summarized in the issue on the agrarian question and globalization of *Alternatives Sud* (Vol. IX, No. 4, 2002).

A new threshold has been crossed with the purchase of gigantic concessions of territory in the continents of the South, particularly in Africa and Asia. Thus a Norwegian company had bought 38,000 hectares from a traditional chief in Ghana. The South Korean company Daewoo had concluded a lease for 99 years of more than a million hectares in Madagascar, half of the arable land in the country: what James Petras has called 'colonialism by invitation'.[6] The Collective for the Defence of Malagasy Land has requested the High Transition Authority to annul the contract, but it has only received an evasive response. Saudi Arabia has set its sights on half a million hectares in Tanzania and China is considering using 2.8 million hectares in Democratic Republic of Congo to develop palm cultivation. In Laos, between 2 and 3 million hectares are the object of such negotiations. In Cambodia, the Chinese Haining Group obtained 21,250 hectares in the province of Kampong Speu. It is reported that concessions for food and agrofuel plantations in Africa have been given to companies in Japan, China and the US.

The FAO talks of a new colonialism. And Jacques Berthelot refers to it as a 'hold-up' of land.

'Green revolution', 'agrarian reform': these terms take on very specific meaning in this framework, destined to promote agrobusiness. The subsidies to farmers in periods of excess supply and which, according to Jean Ziegler, amounted, in 2005, to 349 billion dollars a year, have largely become a means of passing public money to the private sector and to favour those who enter into the logic of agrarian capitalism. And finally, the great unsaid in this whole question is that the transformation of peasant agriculture into productivist agriculture is also – and perhaps above all – one of the new frontiers of capitalism, which makes it possible to offset the crises of accumulation in the industrial and financial sectors. We see confirmation of this in France where, as from 2006, with the rise in the prices of cereals 'investors are all of a sudden interested in agricultural raw materials and speculation is partly responsible for the brutality of the surge in prices'.[7] Demand from certain emerging countries has made it possible to liquidate stocks and the European Union proposes suppressing the 'set-aside' regulations in 2008. The development of agroenergy, also decided by the same EU, acts as an accelerator to the point that some people are posing the question of world food. 'We managed the surpluses so we shall certainly have to manage the deficits', writes Laetitia Claveul. The same thing goes for maize in the Americas and rice in Asia.

It is true that, as Professor Hans Christoph Binswanger, former professor at the University of St Gall in Switzerland, reminds us: 'the added value generated by agriculture is systematically inferior to that of industry'.[8] There are several reasons for this: a demand that is fairly inelastic, as it is always dealing with the same products; more restrictive conditions for increasing production, a slower amortization of the machines used according to the rhythm of the seasons, the limitations of additives, fertilizers and pesticides which are dangerous for health and soil fertility; competition that can only be based on price and small margins between prices and costs. As we shall see, the differences with industry are considerable, not to mention with those that follow the logic of the financial markets. How, therefore, can agriculture become a new frontier for the accumulation of capital?

This can only be done by increasing the demand and to do this there are are three possible mechanisms. The first could be by satisfying quantitatively the food needs of all those among the world population who do not have enough to eat. But that is hardly on the agenda, because the growth model prioritizes the development of 20 per cent of the world's population, ignoring those who contribute only marginally to producing added value and who are unable to become consumers in the short or medium term. The emergence of countries like China and India has an impact on demand because the countries are so densely populated and 20 per cent of the Chinese and the Indians are consuming more bread and meat. This adds up to a good number of clients who are thus capable of contributing, at least indirectly, to using up the European, American and Asian stocks and giving a boost to livestock production in Argentina, Brazil and Colombia. But, for the moment, most of the demand is met by the emerging countries themselves.

The second solution consists of diversifying the products in favour of a qualitative change in demand. In the food sector, this is generally a relatively long process (for example in meat consumption), except when a transport revolution facilitates a new, attractive and innovatory supply. Transported by air, the arrival on Western markets of the new season's fruit and vegetables and exotic products has found outlets among higher- and medium-income sectors of the population. Hence the quasi-industrial development of certain products (i.e. flowers) in peripheral regions, creating considerable ecological damage and social conditions that are often deplorable. This happens in countries that hardly take such considerations into account. It is enough that it brings in foreign currency that can reward local capital and that, thanks to importation, the consumption model of the upper and middle classes is maintained or even increased. However, it should be added that most of the time it is mostly foreign investment that is being used, leaving only a small portion of the rewards to the bourgeoisie serving as intermediaries (*compradores*) but who have influence on national policies.

The third solution is a new, non-food demand, for various types of industries and more precisely these days, for agrofuels which have come just in time to revive the prices of agricultural products and their role as a financial refuge in times of crisis. And this is where the most efficient method of production comes in, where costs are reduced to a minimum and gains are maximized – in other words, monoculture. This is, indeed, transforming enormous areas of peasant agriculture and forests in order to produce a single crop – soya, eucalyptus, oil palm, sugar cane, maize or wheat – with all the ecological and social disadvantages of this kind of agricultural activity. This is why agroenergy is also the new frontier of capitalism, with the twofold advantage of contributing to capital accumulation and apparently responding to ecological concerns that have to be addressed. But it has yet to be proved that the two objectives are compatible.

ECONOMIC AND FINANCIAL ISSUES OF AGROFUELS

The agrofuel sector has therefore created much interest among business circles. As we saw, for a long time, the first reaction was a refusal to admit the causes and consequences of climate change. When the problem was no longer one of externalities but began affecting accumulation, there was a change of view. The approaching peak in the production of fossil fuels and the increase in the price of oil added to climate destruction and led to the interest in agrofuels. Public opinion, alerted by scientific reports, political decisions and mass communications, was ready to legitimize any measures that diminish CO_2 emissions into the atmosphere and to resolve the energy crisis. Besides, several of the great world powers, like the US and the EU, not to mention emerging countries like China and India, were increasingly worried about their dependence for fossil energy on the Middle East and other 'unstable regions' in Africa and Latin America. This encouraged the countries of the North, but also of the South, to promote the renewable sector of energy through a series of measures, from direct subsidies to tax exemption and reduction in customs charges.

The situation was therefore favourable for profitable investments in green energy, especially in agrofuels. In fact, the technologies of the first generation of ethanol and agrodiesel had already been perfected, which made it possible to obtain rapid returns on investment. Research on the second generation (using vegetable waste and then cellulose – wood, in other words) was well advanced, with hopes for quick results. They were also being heavily financed by public funds. And it provided an opportunity to promote new technologies monopolized by powerful groups, notably in the field of GMOs.

There are therefore numerous aspects of the agrofuel sector that attract economic interests. While in the case of oil and gas, public companies have taken back their control, leaving the refining and distribution in private hands – even if certain state enterprises are sometimes dominated by private capital, as in the case of Petrobras in Brazil – agrofuels have entered directly into the private sector, as from the production stage. Certain transnationals and large landowners have acquired enormous amounts of land. In general, it has been local business who, at first sight, seemed to have financial autonomy because they had their own juridical status. But they are often linked among themselves, through common shareholders, institutional or personal, or they form part of more comprehensive groups. This leads them to international capital (when they are not simply the local name of a transnational enterprise) and this is the case both for oil palm and for soya.

Ethanol, in the tropical regions, is usually linked to the large, ancestral sugar estates, belonging to the old oligarchy, now reconverted to agrarian capitalism. Such is the case of Brazil and the Philippines. But there are also international investors, either belonging to the 'hyper-rich' like Bill Gates and George Soros, or automobile, chemical and food industry multinationals. It is therefore important to study the economic and financial issues by looking at the new dimensions of agrobusiness, private and public investments, and the international networks of capital involved in the field of agrofuels.

New Prospects for Agrobusiness

Agroenergy is thus undergoing a boom through agrofuels. It is true that the production and distribution of agricultural products have interested some large multinationals for already more than a century. To these should be added the seed companies, above all since it has become possible to apply genetics to sectors of agriculture and to develop GMOs. The chemical industry has put fertilizer products and pesticides on the market to increase or protect agricultural yields. Links have been created between these different sectors: some companies, like Monsanto, combine several functions.

With the development of agrofuels, two new sectors became interested in agriculture: the oil companies and the automobile industry. For the first, it is a question of conserving monopolies established over energy resources, and for the second, to keep control over new fuels. Some examples can serve to show how many implications and alliances link up around this new activity.

Today almost all oil companies are interested in agrofuels. Total is active in Africa and Shell is investing in research for the production of ethanol from cellulose. BP and Exxon are also involved, as well as younger companies such as Petrobras in Brazil and Repsol in Spain and Latin America, as well as Ecopetrol in Colombia, which possesses 50 per cent of the capital of seven palm-producing companies and has invested 23 million dollars in the Ecodiesel Colombia company. Obviously it is agrobusiness that is most concerned by this activity:

some of them have close ties with oil and automobile companies, not to speak of the agreements they make between themselves. We shall just mention a few to illustrate the current dynamics.

Archer Daniels Midland (ADM), one of the food industry giants, has made an agreement with Cargill and Bunge for producing agrofuels, these two large companies being directly involved in the sector. Thus Cargill, a US company, possesses 2.6 million hectares of transgenic soya. No doubt it is not only to produce agrofuels, but the new trend is to give it priority. The Central Energetica do Vale do Sapucai (Cevasa), in the state of São Paulo in Brazil, serves the interests of several transnationals. Cargill is also building a mega-port to transport the soya grains to Paraguay, with an export capacity of 1 million tonnes a year. A similar development has taken place at Santarém in the Brazilian state of Paraná, through the activities of Cargill Agricola company, its Brazilian branch. Recently the company lost a case against the state of Paraná over a question of fiscal control.

Bunge, also a North American company, has been involved in the exportation of sugar and alcohol from Brazil. For this it bought the Santa Juliana company of Minas Gerais and tried to acquire the Vale do Rosario, the third largest factory producing ethanol in the country. With another giant corporation, DuPont, it has created a local company called Treus, which is to produce hybrid maize and soya and is also engaged with BP in producing ethanol from sugar cane.

The Swiss transnational Syngenta is very active in Latin America, particularly in the development of enzymes for hybrid maize (maize 3.272). It has made a ten-year agreement with Diversa Corporation to produce transgenic enzymes for ethanol.

Monsanto, which with Syngenta and DuPont controls 44 per cent of the sales of seeds in the world, has made agreements for the production of agrofuels based on GMOs, as well as with Dow Chemicals for producing maize seeds that are resistant to eight herbicides, and with BASF corporation for research into new, transgenic forms of maize, soya, cotton and cinnamon for an investment of 1.5 million dollars. Another agreement has been made with Cargill to set up the Renessen company which will also produce transgenic forms of maize and soya for agrofuels and for fodder. These alliances illustrate the aim of Monsanto to build a competing strategy with Syngenta and DuPont.

Other, smaller companies are also present in these sectors. There is Global Food, an American corporation which has allied with the

Santelisa Vale Group in Brazil to constitute the Companhia Nacional de Açúcar e Alcool (CNAA), with an investment of 2 billion reals for the construction of four factories in Goias and Minas Gerais. An example from Africa is the Société française des caoutchoucs (Sofinal), a holding company with its headquarters in Luxembourg, which possesses oil palm plantations in Liberia, Côte d'Ivoire, Cameroon, Nigeria, and in Asia and Indonesia. The Société de caoutchouc de Grand Bereby (SOGB), with help from the International Finance Corporation (IFC), one of the World Bank bodies, has invested 6 million dollars in oil palm plantations in Côte d'Ivoire. The Société camerounaise des Palmeraies (Socapalm), which belongs to the Boloré group in France, is also investing in oil palm plantations in Africa, while in Papua New Guinea, a joint venture, the Pacific Palm Plantation Ltd, has been set up with 20 per cent governmental participation to exploit 23,000 hectares of oil palm.

Chemical and pharmaceutical corporations are also interested in the sector. Bayer, Dow Chemicals, DuPont (which has made agreements with BP for the distribution of ethanol in the UK) as well as BASF. It is interesting to see to what extent industrial strategies are developing from very different points of departure, as we have just seen in all the large traditional sectors. But there are also groups and personalities who see an opportunity for financial gain in this growth sector. This is the case for the Peter Cremer Gruppe, which has invested 20 million dollars in Singapore for a refinery to produce agrodiesel. Other financial enterprises are Kidd and Company in the US (which controls the Coopernavi in Brazil), Merrill Lynch, Stark and Och-Zit Management, investment funds active in Brazil, and Infinity in London, which has invested capital in four factories to produce ethanol in Brazil. From France there is Louis Dreyfus, which has injected capital into the four factories of the Tavares de Melo group of Pernambuco, while Tereos, also French, has invested in Cosan and in the Franco-Brasileira de Açúcar, two Brazilian companies.

Certain personalities, well known in financial circles, are also involved in the sector. George Soros is a shareholder in Adecoagro in Minas Gerais and in Mato Grosso in Brazil and James Wolfensohn, former President of the World Bank, is Administrator of Brenco (Brazil Renewable Energy Company), founded by André Philippe Reichstul, former President of Petrobras, in which he has invested 2 billion dollars. Another shareholder in Brenco is Vinod Khosla of Sun Microsystems, one of the founders of Google. Carlos Slim, the most important Mexican businessman who, according to *Forbes*, has the second largest fortune in the world, has invested in agrofuels in Paraguay. Then there is Bill Gates, shareholder in Pacific Ethanol, a company active in Brazil. In Europe, Nordzücker-Südzücker, and in India, BHI, are also investing in agrofuels. Pension funds begin to be interested, too, and the *Wall Street Journal* of 26 August 2006 warned that there was a real danger that agrofuels could become the object of speculation.

As it is a high-tech sector, with promising future prospects, other kinds of companies are also becoming involved, such as certain forestry enterprises, like Stora Enso, Anacruz, Aranco and Botnia in Latin America. It will be remembered that ethanol of the second generation envisages using cellulose, i.e. a forestry product.

Completely new horizons have opened up finally for the creation of artificial living organisms that can produce energy. The idea is to move beyond purely vegetal sources and to try and enter into new realms of the living, as is being done by Synthetic Genomics in the US.[9]

Collaboration of the Public Authorities

The intervention of public authorities, both national and international, in the field of agrofuel production has already been mentioned. They apply the usual state instruments, like tax exemption for concessions to national and international enterprises as in Papua New Guinea for producing palm oil, or in the US for ethanol based on maize. But they also provide co-financing for research and production.

Thus the Colombian state has given financial aid, in the form of a low-interest loan, to a number of oil palm production companies. This is the case, for example, of Urapalma, which has developed plantations in the region of Chocó, sometimes even illegally, as was shown in a tribunal that took place in the country in 2007. It is interesting to note that support of this kind of plantation has been integrated into Plan Colombia, which in particular finances the use of glyphosate, produced by Monsanto and applied locally by Dyncorp, and which is a herbicide often applied in strong doses and indiscriminately.[10] The application of the Kyoto Protocol is also used to justify the initiative of Monsanto through the Compañia Agrícola Colombiana Ltd, especially for producing transgenic maize, which can supply ethanol using the M810 technology. DuPont of Colombia enjoys similar favours.

USAID has invested 700,000 dollars of anti-drug money to finance palm plantations, within the framework of the Colombian

Agrobusiness Partnership Program. This aims, according to its own definition, at supporting private activities promoting the production or transformation of legal and profitable agricultural products in the regions or areas that are close to illicit cultivation. Enterprises that have benefited from this assistance are mainly in Uraba Union of Palm Growers (Urapalma).[11]

For Papua New Guinea, the Asian Development Bank considers that this kind of production is the best way of reducing poverty and has decided to finance it.[12] As for the World Bank, through the International Finance Corporation which supports the private sector in poor countries, it is financing oil palm plantations in Côte d'Ivoire. The reason given is that it increases jobs, raises living standards, brings in foreign currency and respects the environment. According to the World Rainforest Movement, the reality is quite different, both ecologically and socially.[13]

These examples show, not only the mutual involvement of various economic sectors interested in agrofuels but also the support received from public authorities at both national and international level. Because agrofuels are high tech it is particularly interesting for investors at a time of financial crisis. It has everything going for it: technologies already well honed or in full development, State measures requiring a growing proportion of fuels of vegetal origin and a universal agreement on the necessity to reduce the use of fossil energy.

Moreover, there are constant exchanges between the economic and political systems, making the boundaries between the two very porous, which helps the former to exercise power over the latter. Personalities move systematically between the two. There are gigantic lobbies at work on the parliaments and governments. Thus the political institutions cannot be considered as independent or act as a counterweight to economic power: their autonomy is very relative. It is what is called in the US the 'Revolving Door' effect. This is also applicable to agrofuels.

A systematic study would be most enlightening, but here we can already cite the names of some administrators of enterprises and transnational corporations that have been mentioned in this book and documented by Geoffrey Geuens of the University of Liège.[14] For example, ADM in the US includes among its present and past administrators the following personalities: Brian Malroney, former Prime Minister of Canada; Robert S. Strauss, ex-President of the National Democrat Committee and former US Ambassador to Russia; John R. Block, ex-Secretary for Agriculture in the US government; Richard Burt (Republican), former Ambassador to the Federal Republic of Germany; Andrew Young (Democrat), former US Ambassador to the UN.

As for Unilever, among those who serve or who have served as administrators, are: Lord Brittan, former Vice-president of the EC; Baroness Chalker of Wallasey, Conservative politician and former Secretary of State for International Development; Wim Dik, former Netherlands Minister for Foreign Trade; Lord Simon of Highbury (Labour), former Minister of Trade and Competition in Europe, ex-adviser to Tony Blair, but also Administrator of Suez and Counsellor to the Deutsche Bank, the Allianz and the Morgan Stanley International Bank; Onno Ruding, former Minister of Finance in the Netherlands and Administrator of Robeco Group (Rabobank), of RTL, of Peciney and of Citibank; Claudio X. Gonzalez, former Senator of the PRI in Mexico and former Special Adviser to the Mexican President, but also Administrator of Kellogg's, of General Electric, of Kimberly-Clark; Oscar Fanjul, ex-Secretary-General of the Spanish Ministry of Industry and Energy and Counsellor of BBVA, of Ericsson and of Marsh and McLennan; George J. Mitchell (Democrat), former US Senate Majority Leader, CEO of Walt Disney and Administrator of FedEx, of Xerox and of Staples.

BP had Peter Sutherland as its President in 2007, the former European Commissioner for Competition Policy, former Director General of GATT and then of the WTO, member of the International Advisory Committee of Coca-Cola and President of Goldman Sachs International. As for Royal Dutch/Shell, its Deputy Chairman is Lord Kerr of Kinlochard, who used to head the British Diplomatic Service and was former Ambassador to the United States; Wim Kok, former Prime Minister of the Netherlands; Sir Anthony Acland, former Under-Secretary of State at the Foreign and Commonwealth Office and former Ambassador to the United States.

BASF has had, as Administrator, Viscount Etienne Davignon, former Vice-president of the European Commission and former President of the International Energy Agency and Administrator, among others, of Suez, of Total and of Unicore. Sean O'Keefe, former Administrator of NASA and Deputy Director of the US Office of Management and Budget, has also been a member of DuPont's board of directors, as well as Goran Lindhal, who was Special Adviser to Kofi Annan, Secretary-General of the UN. In Indonesia, the PT Austindo Nusantar Jaya company, which is active in the production of oil palm, has as its Administrators: Arifin Siregar, former Indonesian Minister of Trade and ex-Ambassador to the United States, who has also served as his country's representative at the IMF and the World Bank; Adrianto Machribie, Administrator of Freeport McRohan, a US company associated with this group and for which J. Bennet Johnston, former Republican Senator, is a representative; Roy J. Stapleton, former Secretary of State for Research and Intelligence and former Ambassador to Singapore, to China and to Indonesia, who is also Administrator of ConocoPhillips. And finally, Henry Kissinger, former Democrat Secretary of State also serves as an Administrator of this Indonesian company.

It is also possible to make the list to show how different administrations, governments and other commissions have recruited members of large corporations, such as Condoleezza Rice, former Secretary of State of George W. Bush's administration and previously Administrator of Chevron, and Kathleen B. Cooper, former chief economist with Exxon, who became Under-Secretary for Economic Affairs at the US Department of Commerce.

The aim of citing all these names is not to promote a conspiracy theory or to question the integrity of the various individuals but to show how a certain logic leads finally to a system that removes differences and tends to unify interests. It is thus difficult to establish the various competences and to believe in the independence of political decision-making. The counterweights are weak compared with the power of the economic interests and their logic.

AGROFUELS AND THE FOOD CRISIS

Recently there have been many cries of alarm about the world food crisis which, as we know, is essentially due to the lack of means with which to obtain food and not to the lack of land to produce and feed its population. As we have already said, the FAO considers that agricultural production can ensure the subsistence of some 12 billion people, while UN demographic forecasts are for a world population of 9,300 million in 2050, also proving that the land has the necessary resources to feed so many inhabitants in the world. Nevertheless, the World Bank states that needs will double between now and 2050.

According to the former Special Rapporteur Jean Ziegler, in 2008 there were 854 million people suffering from hunger because of poverty and 2 billion others who were malnourished. FAO Director-General Jacques Diouf, at a conference at the University of Havana in July 2008, declared that there was an increase of 50 million people suffering from hunger in the year 2007. In the period

2007–08, more than 115 million people fell below the poverty threshold. There is little doubt that the increase in the price of food has had a direct effect on the world hunger phenomenon. Apart from sugar, all food products have increased in price since the beginning of the twenty-first century, with a steep increase as from 2007. This has led to food riots in various countries across the globe, as in Haiti and African countries like Senegal, Burkina Faso and Egypt, as well as to social disturbances in other parts of the world. The impact was evidently much stronger on the poor countries that depend almost entirely on imports for their food. Thus, according to the FAO and cited by Laetitia Chavreul,[15] the imports of cereals, the price of which have skyrocketed, increased the cost of the food basket by 90 per cent in developing countries, as against only 22 per cent in the richer countries. At the same time, the powerful food industry had record profits. Cargill declared that its profits were up 86 per cent for the first quarter of 2008, compared with that of 2007. According to Jacques Berthelot, in 2007, 100 million tonnes of grain have been transformed into agrofuels, which would be enough to feed 450 million people: nearly half of the 963 million people who were undernourished in 2008.

To what extent has the development of agrofuels had an influence on the increase in the price of food? Partisans of agrofuels based on agriculture affirmed that other factors were involved, while adversaries claimed that, on the contrary, they played a central role. To clarify this problem we refer to the excellent study by the French economist Jacques Berthelot, specialist in agrarian issues, who has specifically dealt with the explosion of agricultural prices.[16]

As the author points out, there are two kinds of basic reasons that are affecting the increase in the price of cereals: those concerning the supply, and those linked to the demand. For the latter, he cites: the increased production of agrofuels which, in certain regions, reduces the volume of cereals available for food, both for humans and for animals; the increase in food consumption in the emerging countries, like China, India and Brazil; the world demographic increase; and speculation in sectors of the economy that have taken over from other sectors currently in crisis. As for supply, three factors are mentioned: the drop in production caused especially by changes in climate, the increase in the price of oil which affects the cost of inputs such as fertilizer and pesticides, as well as of transport, and restrictions on exports by countries who want to ensure their food security.

Berthelot adds that it is of course necessary to distinguish between the causes that are structural and those that are only temporary. The former include climate change, lack of investment in the agricultural sector and the gradual change in diets in the emerging countries, whereas the latter include the bad climate conditions (notably in Australia in 2007), reduction in stocks and the sudden development of agrofuels – a factor which obviously may become structural.

The World Bank's response to the question was conclusive. According to the calculations of its experts (especially Don Mitchell), agrofuels are responsible for 75 per cent of the increase in price of food products. In the field of maize specifically, according to the IMF, agrofuels were responsible for 70 per cent of the rise in food prices, not to mention the indirect effects that such an increase can have on other cereals, as US farmers abandon other crops in favour of maize.

In this field, the US was largely responsible, given the size of its production. An increase in the price of maize because of agrofuels has repercussions on the world price of the commodity, with consequent food insecurity for populations which depend on it as their staple. Mexico, in particular, suffered bitterly from this experience. In fact, following its participation in the North American Free Trade Agreement (NAFTA) with Canada and the US, Mexico has become a large importer of maize. The soaring price was all the more remarkable because the production of ethanol from maize had been considered an economic heresy which could only be maintained with state subsidies. The real reasons of the situation are political, above all the desire to reduce the energy dependence of the US. However, there were also economic reasons because of the power of the food industry. The production of maize rose from 607 million tonnes in 2006–07 to 776 million tonnes in 2007–08, mostly because of the general craze for producing ethanol.

In Europe the pressure was also strong, because to attain the objective of 10 per cent of agrofuels by 2010, which was to be mixed with fossil fuel for automobiles (this aim is now being questioned even within the EC), some 20 per cent of arable land would have to be converted to agrofuel crops. Europe does not have enough land for this and hence its interest in the developing countries that have enough arable land to be able to cultivate such crops and thus meet Europe's needs. It is true that in 2008 the prices suddenly dropped, but they are still above the prices of the beginning of the decade.

It is thus clear that although agrofuels are not the only cause for the increase in food prices, they play an important role. True, in certain countries, as in Brazil, disposable land makes it possible, theoretically, to combine food production together with the development of agrofuels. However, in actual practice the situation is very different. Added to the monoculture of soya and eucalyptus, sugar cane cultivation has displaced certain food crops and cattle ranching, which in turn has repercussions on the reduction of forest land. In other regions of the world, above all in the South, the consequences of extending monocultures has a direct impact on the disappearance of forests. As Jacques Berthelot showed in his study, India and China were hardly a determining factor in the increase of food prices because, up until now, these two countries have themselves satisfied the increased food needs of their population.

In light of the accumulation of the contradictions and insignificant effects of the increase in agrofuel production, it is important to find out the reasons why they are being so intensively developed.

AGROFUELS AND THE REPRODUCTION OF CAPITAL

There are various ways of tackling the problem of agrofuels, both from the technical viewpoint and from the consumer angle. In this book we want to put the emphasis on the economic function of their production and in particular the role that they play in the reproduction of capital in this time of financial and production crises. It would seem that the sudden increase in the production of agrofuels is caused by the logic of capitalism and that this in turn explains the sudden and rapid development of a very precise sector of the economic system, that of energy, which is strategic for all human activities. This is not a very surprising discovery given the hegemony of the capitalist logic over the rest of the world economy. However, the question is all the more central because the intense use of energy is at the very centre of the development model promoted by capitalism, at a time when fossil energy is reaching its peak and the general awareness of climate change has become a political problem.

In fact, as we have seen all along while carrying out this work, there is a twofold problem that affects fossil fuel. First, the gradual exhaustion of the sources of energy, particularly oil, gas and in the longer term, coal and now, even the mineral source of uranium. For some of these, particularly oil and gas, we have already reached the peak of our reserves, or gone beyond it. This is particularly the case of the US and certain European countries. Hence the problem of dependence on other regions of the world for the supply of fossil fuels. This constitutes the first problem for those responsible for the capitalist economy. Without new energy resources and, in the immediate future, without control over current energy resources, the system cannot reproduce itself. Hence the need, first to ensure the supply from regions that are politically and militarily controlled by the West and then, to move rapidly beyond the cycle of fossil energy. This is how agrofuels take on their significance as an integral part of renewable energies. It is clearly a genuine and objective question to which there are several responses: finding solutions that meet the requirement of their exchange value, the basis of the capitalist system or, on the contrary, being concerned with their use value, that is to say, the satisfaction of the 'well-being of peoples'. We shall return to this issue in our conclusions.

The second problem is that of the climate damage provoked by fossil fuels and the urgency of taking measures to save the environment. Capitalism considered ecological problems as externalities until the day came when the gravity of the situation began to affect the accumulation process. This is a relatively recent development, but it demands attention and has to be taken into consideration, even within the logic of capital. Here, too, various responses are possible, those that take into account the whole energy cycle of agrofuels and their effects or those that only consider the immediate factor of their technical application at the level of energy production. Before tackling the question of the reproduction of capital in detail we should consider the role of agrofuels in the new energy policies.

Role of Agrofuels in the New Policies

Agrofuels are above all used in transport. But this sector only represents 14 per cent of greenhouse gas emissions (10 per cent from road transport, 2 per cent from maritime transport and 2 per cent from air transport, with the latter two rapidly increasing). Industry is responsible for 31 per cent and the residential sector and services for 19 per cent. For the latter activities it is above all a question of using diesel, but also electricity supplied to a large extent by coal and nuclear energy.

Agriculture is responsible for some 18 per cent of greenhouse emissions, particularly livestock production, which emits 70-75 million tonnes of gas a year. It should be remembered that meat consumption, above all because of the diet changes in the emerging countries, could double by 2050.

Transport, therefore, only accounts for a relatively modest share of the problem affecting the climate – even though it is not negligible. We should remember that in Europe and the US, the aim is that the share of renewable energies in transport should be 20 per cent by 2020. This means that 80 per cent of the energy will still be of fossil origin. If half the European arable land were to be converted to producing energy from vegetal sources, this would only partially meet predicted needs and probably hardly cover the increase in demand between now and 2020. Besides, the whole process of production and distribution of agrofuels does not resolve the climate problem. Their yield is relatively smaller than fossil fuels and so more must be produced to obtain the same result. Furthermore, directly or indirectly, CO_2 emissions are hardly inferior to those from fossil energy, if one takes into consideration the whole cycle of production, transformation and distribution. Even if these factors can be improved by technological advances, it is very clear that they play only a minor role in the solution. Hence, once again, why is capital so interested in agrofuels these days?

Agrofuels in the Reproduction of the Capitalist Economic System in the Short and Medium Term

As we have shown, the reaction of the neoliberal discourse, when faced with the climate problem, was at first to deny the existence of the problem, or to minimize it. There was a delegitimization exercise and contempt for the scientific discourse of climatologists. As in many other cases in the history of capitalism, the reaction was then to transform the problem into an opportunity, which ended up optimistically: science and technology would succeed in finding solutions. Finally came the turn of the ecological discourse and even what could be called the 'green lies', showing to what extent recent industrial technologies, the improvements in the consumption of vehicles and even the new mixtures of petrol and diesel contribute to protecting nature and improving the climate. In reality they are prolonging a consumption that is destroying the environment, even if it is on a reduced scale, and a large part of these advantages is absorbed by the increase in energy requirements.

Moreover, the first reaction to the reduction in the sources of fossil energy was to exploit new reserves, particularly in biodiversity zones, and then to exploit other sources of energy production, such as nuclear plants and agrofuels – all in order to maintain current modes of consumption. During the last few years the intensive research for new frontiers for capital accumulation has become even more vigorous, mainly because of the production and financial crises and the lowering of the rate of profit. Public services have been increasingly privatized, enabling them to be turned into commodities and peasant agriculture is being transformed into capitalist enterprises.

Four groups of transnational corporations are now working together: the oil, chemical, agrobusiness and automobile groups. This enables them to conserve or recover control of the production and distribution of energy in the new fields and hence to increase their source of accumulation. It is in no way a plot, but simply the result of the very logic of an economic system that favours oligopoly, which is what we would like to explain in greater detail. There are many factors that help to make agrofuels an important element in capital accumulation.

Control of landownership, the base of production

As the origin of ethanol and agrodiesel is vegetable-based, the control of land, both direct and indirect, is necessary to bring the sector under control. The first step consists of acquiring land, and Brazil is a good example. The second is achieved by controlling the work of small peasants, who keep their land but who, through contracts, become dependent on the large agrobusiness corporations. This is what Monsanto does, particularly with the introduction of genetically modified organisms, but also with the oil palm companies. The subordination of the work of peasants to capital is carried out in various ways: supplying seeds and plants, selling fertilizers and pesticides, purchasing the product and sometimes giving loans at usurious rates to indebted peasant families. As for the price offered, it barely covers the costs of the reproduction of the populations concerned. It is what some people have called 'the capture of the small producers'. In this way, one of the bases of the capitalist system is constituted: the control of the elements necessary for production, either by the direct acquisition of property or by the total submission of the worker, who remains the owner of the land.

Labour exploitation

It is mainly in the South that the level of exploitation of workers remains physically extreme. In certain cases, it is a form of slavery, as in many sugar plantations in Brazil and Colombia. In other cases the work is paid at a minimum level, virtually without social security or pension. Unions are forbidden or rendered ineffectual by repression or corruption. When they are small peasants, remuneration is barely above subsistence level and conditions are similar to those of serfdom. As for monoculture, it eliminates much of the work, causing migration towards the towns and making it possible to reduce the use of labour to a strict minimum. In other words, the maximum exploitation of workers is widespread and corresponds to the logic of capital, which consists of exercising pressure on the various elements involved in the costs of production in order to maximize the profits and thus appropriate the added value. The social costs of such operations are not included in the capitalist accountancy and have to be supported by the collectivity and by individuals.

New technologies that are ecologically very risky but highly profitable

These technologies are the promotion of GMOs, particularly for soya and maize, and they will also be for trees the day that technology makes it possible to use them for producing energy of vegetable origin. The risks of GMOs are well known. Although they improve the profitability of plants and animals, their extension into forms of monoculture risks endangering numerous species, the long-term effects of which have not really been calculated. This activity is dominated by a few chemical giants and agrobusiness: Monsanto, Cargill, Bunge, Bayer, etc. Many governments have imposed brakes on the use of genetic modifications but very often their efforts have been blocked, both by the power of the transnational corporations and by the fact that the seeds are transported from one region to another by the wind, insects and other animals, it being impossible to control. It should be noted that this form of culture requires a lot of water and certain collateral effects, particularly on soils, are not negligible. The increase in productivity, thanks to the new technologies at a time of rising prices, obviously makes it possible to increase profits, which of course forms part of the capital accumulation process.

Exclusion of externalities

As we have already said about peasant migrations to the towns, a series of costs are thus transferred to the collectivity or placed upon individuals. The destruction of biodiversity and carbon sinks, water pollution, soil contamination, the sterilization of the seas – all these add up to heavy costs which are not taken into account as long as they do not affect the reproduction of capital. It is the collectivities that have to bear the effects of such practices in the medium and the long term. The enormous extension of monoculture ends by creating desertification, depleting the soils, diminishing underground water and destroying biodiversity. These natural disasters are the direct result of the elimination of such costs from the accounts of productive operations. But one day, we are all going to suffer from the effects, including financiers. It is up to the countries to take action, to the extent they are capable of it or, quite simply, the well-being of citizens will be severely compromised.

As we have said, the expulsion of peasants from their land is bringing about uncontrolled urbanization in most of the countries of the South. This is caused not by the development of urban or industrial activities but by a demographic surplus from the rural areas, due not only to a population increase but above all to the systematic expulsion of the small producers through the development of capitalist agriculture. The effects of forced migrations all over the world are well known. The wall built on the frontier of Mexico with the US, which prevents impoverished peasants from emigrating to the North, is an illustration of this phenomenon. Each year this wall causes more deaths than during the whole existence of the Berlin Wall, while in Europe bodies of dead Africans wash up on the shores of Italy and southern Spain. The causes can be different from one place to another: the systematic concentration of landholdings, or the effects of climate change, but the result is the same. Whole populations are paying the price of the logic of economic development, for which these factors are 'external' to economic calculations.

Furthermore, the individualization of responsibilities is another characteristic of neoliberal thought and practice. The migrations towards towns or foreign lands are attributed to personal decisions. Thus, in Colombia, President Uribe believes that the problem of displaced people (internal migration, mostly because of the concentration of land) must be resolved case by case through administrative decisions, and not judicial ones, whereas the problem is created by a social logic. This makes it possible to hide the real responsibilities and prevents it from becoming a legal problem, confirming, definitively, the expropriation of land. The individualization of the problem thus becomes a mechanism of externality.

Transfer of public funds to private interests

The production of agrofuels not being profitable at the moment, even in a period of increased oil prices, production subsidies, exemption of sales tax and the lowering of customs tariffs are required to make them competitive with fossil energy. It is true that the rapid increase in the price of oil and gas has somewhat reduced the importance of such transfers. But in most cases the need for subsidization by public funds is still important. An analysis of the accounts shows that it is the big companies that take most of the assistance from the state. Agrofuels thus reproduce the classic mechanism that is to be found in other fields, particularly in armaments and subsidies to agriculture, both in the US and in Europe. It is what could be called 'green neo-Keynesianism'.

The logic is always the same. To the extent that some sectors are not profitable or in the case of financial crisis, capital calls on the state which uses public money to encourage risky investments or to save financial capital from disaster. Public funds are the common heritage that, on the pretext of economic efficiency, is then transformed into private assets, the source of accumulation.

Public investments for the infrastructure required for new industry – roads, railways, ports, warehouses – are necessary not only for the production but also for the distribution of agrofuels. Considerable outlays are being made in these fields, which are often then privatized.

Reproduction of North/South dependency

The main requirements for new sources of energy are in the countries of the North, to which we can now add the emerging countries like China and India. Most of the production of agrofuels is, however, in the South, which is forced to bear the ecological and social costs of such operations. Ethanol is promoted in Brazil, Ecuador, Argentina, Philippines, while agrodiesel is more common in regions or countries like Colombia, Malaysia, Indonesia, Papua New Guinea and central Africa. For the time being the refineries are built in the North, as in the Asturias in Spain and in the US.

So the model of economic dependence is reproduced once again, the capitals of the North being where the decisions are made. No account is taken of the externalities in the South, until they affect the rate of profits. Even if the production of agrofuels is a source of foreign currency for some countries, this does not mean automatically that there will be an autonomous development, or a benefit that is socially shared. In fact, agricultural rent, like that of oil and mines, encourages the formation of a local elitist social class, which is essentially oriented abroad and which plays an intermediary role between the transnational corporations and the local population. It is not really a national bourgeoisie, ready to invest in production initiatives that first target their own population.

The model is essentially export-oriented. The interests of this local elite tend to increase the exports, a source of foreign currency, of which a large part serves to increase imports that provide them with a standard of living that is comparable, if not superior, to that of the same social classes in the industrialized countries. Thus an upper middle class can enjoy goods brought from outside while most of the population, above all the poorest, see no real improvement in their standard of living. As according to the neoliberal logic the state has been amputated of numerous functions, notably the redistribution of wealth, social inequalities continue to grow.

All that is part of the very logic of capitalism which has every advantage in developing intensively the purchasing power of a minority that can obtain assets of high added value, rather than produce ordinary goods for the majority of the population.

Speculation

The forecast of a rise in prices for agricultural products, because of the increase in demand, due partially to agrofuels, immediately let loose a series of speculative practices. Finance capital thus increased its gains, with pension funds not being uninvolved. The support of agrofuels for the virtual economy is thus considerable. At the beginning of the millennium, there was a move of speculation from oil and other sectors of the economy towards food products. While there have been various factors responsible for this, it turns out that agrofuels was one of them, and not the least important.

The re-legitimization of capitalism

A new discourse appeared, that of the green economy. Almost all the production and distribution sectors fell into step. A new hegemony was built on the consensus of public opinion which is now very sensitive to problems of climate change. The economic actors are presented as benefactors to humanity, which gives them a new legitimacy.

It is true that industry has made some efforts to diminish CO_2 emissions and, in other sectors, to avoid waste. This has had a real and beneficial effect. However, that is not the whole story. In fact, industry's efforts were made at the time when the emission of greenhouse gas had become a real cost and were no longer just an externality. It was therefore necessary to take action to try to reduce the cost of ecological impasses, as well as on the measures taken by governments to apply the Kyoto Protocol. The measures against pollution having become current practice, it cost less to take the initiative to diminish noxious emissions than to continue to pollute.

Relegitimization involved a massive recourse by industry to publicity and thus to social communications. Huge sums were spent on billboards praising the ecological nature of corporations and their products, and to finance announcements to the press, radio and television. That is the price of consensus. The mass media depends for its survival on such financial support which, one way or another, reduces their critical capacity concerning the content of publicity. All that forms part of the logic of capital, which needs to have an ideological basis, without always worrying too much about any dissonance between the publicity and reality. This has been attested by various condemnations of Monsanto and several other firms for the havoc they cause on the environment.

To conclude, it could be said that the function of the development of agrofuels is indeed that of quick profits, a sure source of accumulation in the short term. But they are in fact in no way, or only slightly, a solution to the climate problem and only marginally so for energy consumption. Only massive production, covering hundreds of millions of hectares could make a substantial contribution to the energy crisis and one can hope that popular and political resistance will not permit it. The ecological damage and negative social effects are considerable. The economic benefits affect only certain sectors of capital interests, but as they have numerous ramifications in strategic areas, there is an overall benefit for the world capitalist economy as a whole. Speculation is a good illustration of this. The main function of the industrial development of agrofuels is thus the reproduction and accumulation of capital in the short and medium term.

AGROFUELS AND THE DEVELOPMENT MODEL

We must take one more look, if we are to gain a more in-depth perspective, at the development model itself. True, the great advances in the use of energy, thanks to fossil fuels – coal, first, and then oil and gas – have built the material bases for the utopia of the Enlightenment Century. Such a utopia consisted of a linear progress of humanity towards a limitless future in which the human species would consolidate its mastery over nature. Science was to play a key role in this and its technological applications made it possible to multiply these possibilities considerably, thanks to the new energy resources.

The capitalist economic system detached production from the worker through the division of labour and industrialization, making capital the dynamic engine of economic activity. This made it possible to achieve rapid progress in providing goods and services, integrating an increasingly massive use of cheap energy as a decisive factor, not only in production, but also in distribution. As a result we have a development model that has pushed aside or marginalized all other forms of production and which owes its legitimacy to its own success, making it possible to establish its logic as evidence.

The massive move to using oil as from the mid twentieth century considerably increased work productivity and was responsible for even greater velocity in the production and distribution of goods and services. It was also at the origin of the development of industrial agriculture and enabled the growth of financial capital by exploding the monetary sphere and creating bank money. It also transformed military methods and ways of waging wars.

Because of the major role played by energy, with the double crises of fossil energy resources which are becoming exhausted and the climate destruction which is linked to them, the question is posed of the development model itself, with all its components, material conditions, social repercussions and consumption modes. The problem is to know for how long humanity can conceive capitalist development as its only future when its contradictions are accumulating and the challenges becoming more specific.

One of the obstacles to new solutions lies in the importance of economic issues, which blind the leading players, who are above all concerned to reproduce a system that gives them a dominating material, political and cultural position. Having thus internalized the model, they identify their interests with the well-being of humanity. Agrofuels, the relative utility of which we do not question, provide the pretext for their headlong rush into this field, for they seem to be the solution that will make it possible to reproduce the same model thanks to these new technologies.

While the energy and climate crises seem increasingly to mark the end of a model, it is continuity that fills their imaginary. Solutions are sought that in no way change power relationships concerning economic decisions: neither how to produce, nor the way in which world wealth is distributed, nor the mode of consumption. However, there is every indication that the current pace in the use of energy cannot be maintained and that the new energies will contribute only modestly to the expanding energy needs, such as they are forecast.

Moreover, the economic, social, environmental and political costs of new solutions are very heavy and are questioning their real effectiveness in improving the climate and in meeting energy requirements. Certainly, we do not want to be absolutely pessimistic, nor do we believe, as some do, that the date for the end of humanity is already on the agenda. But we must recognize realities. On the one hand, the climate and energy crises signify the end of the illusion of limitless growth and, on the other, the inequalities in the living conditions that have been created at the world level have ended in a social system that is economically and morally insupportable.

The logic of capitalism is incapable of responding to the challenge, as it continues to consider as externalities everything that does not enter directly in the calculation of exchange value. The mode of production and distribution linked to this logic is not sustainable, as it is based on an overuse of primary materials and energy through the manufacture of disposable products, or ones with a short life, by the extension of transport due to delocalization and the dispersion of the sites of production, and by the liberalization of trade that gives advantage to the strongest. As for consumption, it is established within a model that has been brought about by the logic of accumulation, therefore by exchange value rather than by use value.

Indeed, contrary to what is generally believed, that the customer is king and that demand conditions supply, the opposite prevails. Consumption is conditioned in its practices by the structure of economic production and by all the ideological apparatus that supports it, both to legitimize and stimulate the purchase of goods and services offered by the market. Luckily, science and technology will produce a certain number of concrete responses. There will no doubt be some important progress in the next few years in economizing energy and in the use of new energy sources. Cheap fossil fuel, the cycle of which has come to an end, had not encouraged research and investments in this field. Today there are many projects and experiments are flourishing. But that is not the fundamental problem. It is the whole philosophy of development that is at issue.

Some talk of 'degrowth' – a term that was already used by the Club of Rome during the 1960s and was taken up again by Ramón Fernando Durán, member of the Ecologists in Action, in his remarkable work on oil (*La historia trágica del petróleo en el mundo*), written for the World Petroleum Congress that took place in Madrid in June 2008. The author, observing the end of the fossil energy cycle and the illusion of an energy transition while conserving the model of current growth and accumulation, concluded that the pursuit of the neoliberal project leads to the deepening of the crisis and to an inevitable way out through warfare. Hence it is a radical change that is becoming necessary and what Durán calls 'the post fossil transition through de-growth', which would also benefit the climate. Rather than speak of 'de-growth', a difficult concept for a public opinion influenced by contemporary consumerism, we prefer to talk of the substitution of quantitative growth by qualitative growth. It is not to diminish well-being, but on the contrary to improve it through a better quality of life. We'll come to that in our conclusions.

Resistance to the presentation of agrofuels as a response to the double contemporary crises has not been long in coming. Awareness of the social and individual costs of the externalities of the model has developed rapidly, both at the ecological and social levels. Many peasant, ecological and worker movements have been active and, gradually, so have the authorities. They have put forward the arguments we have used in this book to show the limits of using agrofuels as a way of combating climate change and also to denounce the negative effects on the natural environment and peasant populations. These reactions have also reached governmental level, particularly in Europe, where a brake has been put on the first frenzies of this technology in order to make the governments of member states of the EU moderate their enthusiasm.

Even the most radical movements, like the Brazilian Landless Workers' Movement, do not exclude the use of agrofuels completely. This was confirmed by the Cramer Commission in the Netherlands in 2006 and by the UN Energy Report in 2007. It is clear to everyone that the cycle of fossil fuels is coming to an end and that their negative effects on the environment are most harmful. So other solutions must be found. The utilization of agrofuels has a role to play but it is much less decisive than was at first thought. The conditions for accepting the production of agrofuels by the ecological and social movements can be summarized as follows:

- respecting biodiversity, that is, renouncing monocultures in favour of diversified plantations that do not endanger the existing plant and animal species;
- limiting the agricultural frontier, that is, avoiding encroachment on forest land, especially primary forests. This means using land that is available, and legal protection for carbon sinks and

biodiversity zones, as well as for the settlements of indigenous populations;
- respecting soils and underground water, which means prohibiting the massive use of chemical fertilizers and pesticides and favouring organic agriculture;
- promoting peasant agriculture and helping to improve its working methods, access to credit and the marketing of its products;
- combating the monopoly of the transnational corporations.

If these conditions were met, the production of agrofuels would automatically be oriented towards the needs of local populations. It is indeed possible to respond to such needs by production that respects these five principles. But it is clear that this means a radical rejection of capital logic and subordinating the economy to basic human needs. Solutions, like those of the so-called second generation, could doubtless increase the role of agrofuels in solving energy and climate problems, but they must remain modest in their claims for the future. The possibility of producing surplus energy for urban populations and a certain amount of land for a more intense cultivation for collective requirements is an issue that evidently still remains open, but such activities must have precise ecological and social limits. No global solution will be found without challenging the contemporary development model and posing the question of alternatives.

9
Alternative Ways of Solving the Climate and Energy Crises and the Role of Agrofuels

Is there a solution to what today seems like a situation without a genuine response and, if so, what will be the role of agrofuels? Will the new technologies make it possible to support efforts to reduce consumption? On what conditions can the definition of needs change? We shall now try to respond to these questions.

SOLUTIONS ENVISAGED AND THEIR LIMITATIONS

Three main paths are envisaged to solve the double crisis about which we have talked so much in previous pages: reduction in the consumption of fossil energy; the use of non-agricultural renewable sources of energy thanks to technological progress; and recourse to agrofuels.

Reduction in the Consumption of Fossil Energies

As this issue is widely discussed these days because of vigorous campaigns, we shall just give a brief outline of its main thrusts. Among the initiatives under way that seek the reduction of fossil energy consumption, these are – apart from the efforts being made by industry to reduce CO_2 emissions – a more intense utilization of public transport, reduction in thermal losses in houses and designing new ones that best capture the sun's rays. Then there are the low-consumption light bulbs, heat pumps both to limit the use of chlorofluoros and heating with gas or fuel oil for supplying hot water and air conditioning. Along these lines some 15 cities, including New York, Chicago, Tokyo, Toronto and Karachi, have signed an ambitious programme valued at 5 billion dollars to modernize old buildings, making them more efficient in energy.[1] Many towns and other local administrations are preparing climate plans.

Non-agricultural Energies and the New Technologies

Let us remember that renewable energy means energy that has a natural cycle which is constantly renewing itself and is not used up by immediate consumption. Among the non-agricultural sources these are: hydroelectric energy, solar energy, wind energy, waste energy, hydrogen battery energy and geothermal energy. We shall not enter into all their technical details, but rather concentrate on the extent to which they can satisfy energy needs and their ecological effectiveness. In fact, as Yves Scania and Nicolas Chevossus have written, 'renewable energies suffer from serious defects. First they are not "green" 100 per cent. Second – and this is original sin and it affects them all – they have weak "energy density".'[2] According to Greenpeace, only a third of energy consumption by 2030 will come from renewable sources: 19 per cent from biomass, 11 per cent of hydraulic origin and 3 per cent from other sources.

Hydraulic energy

Hydraulic energy and hydroelectricity are obtained by their conversion from different flows of water (larger and smaller rivers, waterfalls, sea tides, etc.). The kinetic energy of the water flow is transformed into mechanical energy by a turbine, then into electric energy by an alternator generator. Hydroelectricity is a renewable energy and it is considered as an energy in itself, although some environmentalists contest this because of its impact on land and, more recently, because of its carbon balance sheet.

The construction of dams on rivers to turn turbines only really got going after the 1920s. Half of those that existed at the beginning of the twenty-first century were built between 1920 and 1975 and the other half between 1975 and 2000. In other words, the neoliberal era saw their numbers increase significantly. After this date, and more precisely after the 1990s, there was a drop in construction. In fact, the technical difficulties and popular resistance increased and water began to decrease.[3] Furthermore, the sites of the dams were often far away from consumers, which meant very long high-tension lines were necessary.

In 2004, hydroenergy represented 16 per cent of the world electricity production. In France it was equivalent to 92 per cent of all the renewable energies in the country. However, the recent history of big dams has become more and more controversial in China, India, Brazil and South Africa. The ecological damage brought about by their construction has earned them increasingly severe criticism. The Aswan Dam in Egypt, for example, has caused the erosion of the Nile delta.

As for the demonstrations against dams, it is difficult to count them there are so many, among the displaced populations, the indigenous peoples expelled from their land, the villagers threatened by flooding. This was the case in China *apropos* the Pubugou dam. But the most spectacular example in that country is the Three Gorges dam on the river Yangtze. This is supposed to generate 18,200 MW of electricity and the whole project stretches for 600 kilometres. One hundred dams are scheduled for construction by 2025 in the northern course of the river, but the projects have been seriously undermined by the severe droughts that affected the region in 2007 and 2008. It is necessary to flood 1,000 cubic kilometres and displace 1.3 million people for the whole project. Their fate is not enviable. As for the Xilnou dam, the second most important one, it is supposed to become operational in 2015, but it too has come up against resistance. In Latin America, dams are the object of many conflicts, especially in Brazil around the Rio San Francisco.

It is possible to conclude that hydroenergy based on the construction of big dams has reached its peak and that it will no longer be seen as a solution apart from those that already exist. In contrast, many small-scale works create only limited damage and serve to provide water for irrigation as well as a source of energy for specific regions so they can respond to numerous needs. Only small-scale hydraulic energy seems to have a certain future, above all in the countries of the South, with the aim of supplying local communities.

There have also been experiments in using the oceans. Tidal waves as an energy source was already known about in the twelfth century, but it is at Rance in France, since 1996, that the principle has been applied on an industrial scale. The UK has invested dozens of millions of euros in marine energy programmes. However, it is difficult to make them profitable. Another technology is to capture marine energy by using the movement of the waves. There was an experiment in this in the island of Reunion, but the installations were swept away by a cyclone. In neither of these two cases can results meet the challenges of world energy.

Solar energy and photovoltaic cells

Solar energy is at the origin of the main existing energies and it constitutes the greatest hope for the future. We are speaking specifically of the solar photovoltaic cells or captors, while remembering that it is also photosynthesis that is at the base of the production of biomass used as a source of green energy.

Photovoltaic cells are semi-conductors capable of converting light direct into electricity. The phenomenon was discovered by Antoine Becquerel in 1839, but nearly a century had to pass before scientists explored it further and could exploit it. Photovoltaic technology is now in full expansion. In all four corners of the world, various possibilities have been studied and then experimented with in the hope of future commercialization.

Because they are complex and their production yield low, the manufacturing of photovoltaic modules has been costly which has put a brake on mass sales. It is to be hoped however that photovoltaic technology will become more mature, with simplified procedures and better production and that an increase in the volume of production will reduce the cost of the modules.

In fact, in spite of these difficulties the evolution of photovoltaic technology is globally positive. The methods of manufactures have been improved. Currently, 90 per cent of the total production of modules is in Japan, US and Europe, through large corporations such as Siemens, Sanyo, Kyocera, Solarex and BP Solar (these companies alone command 50 per cent of the world market). The remaining 10 per cent of the production is supplied by Brazil, India and China, the chief producers of modules in the developing countries.

In the US, an ambitious plan called 'One million solar roofs by 2017' has been launched in the state of California.[4] The aim is to construct buildings (or adapt them) using non-toxic material, with zero pollution and with electricity that will come from solar energy. There are other big projects proposed, like the installation in the Sahara of a field of solar panels covering several square kilometres which could supply Europe with energy, but even if this is technically feasible it poses political problems that make it difficult to carry out and, in any case, is vulnerable. This is not preventing Germany from developing bold projects. A consortium has been set up to study solar panel projects in Algeria and other countries of North Africa, as well as columns of water that change into steam and activate turbines supplying electricity. This would be transported across the deserts of North Africa to Europe by cable. The EU is also interested in such projects.

Solar energy does seem to be a genuine solution in time, mainly for domestic purposes and for local transport. However, if it is to be generally adopted the price of production of voltaic watts must still be divided by two or three, believes Jean Pierre Joly, Director of the National Institute for Solar Energy (INES) in Paris. But there are real technological advances. Researchers at the Federal Polytechnic (EPFL) in Lausanne and Stanford University in California have succeeded in adding more colours to the colour system of solar cells (GROTZEL), thus increasing their yield.

Composition of a photovoltaic cell

The photovoltaic cell is composed of semi-conductor material that absorbs light energy and transforms it direct into electrical current. The functioning principle of this cell uses the properties of [light] rays and those of semi-conductors.

The individual cell, the basic unit of a photovoltaic system produces only very weak electric power, from 1 to 3 watts, with a tension of less than 1 volt. To produce more power, the cells are assembled to form a module (panel). The connections in a series of several cells increase the tension for the same current, while putting them in parallel lines increases the current and conserves the tension. Most of the commercialized modules are composed of 36 cells of crystalline silicon, connected in series for the application of 12 volts.

The interconnection of the modules – either in series or in parallel – to obtain still more power defines the notion of a photovoltaic field. The photovoltaic generator is composed of a field of modules and a collection of elements that adapt the electricity produced by the modules to the specification of the receivers. These elements, also called the Balance of System (BOS), constitute the equipment between the field of the modules and the final charge, i.e. the rigid (fixed or mobile) structure on which the modules are placed, the cabling, the battery in case of storage and its charge regulator, as well as the inverter when the system uses alternative current.

In 2008, a vehicle running on solar energy did a tour round the world. And it is interesting to note that some 60 projects have been identified for a 'Solar Plan' in the framework of the Union for the Mediterranean,[5] but according to Peter Thiels, Vice-president of the Sharp Corporation's Solar Energy Department, the bank crisis has seriously reduced investments in this field.[6] Nevertheless research continues: for example, one recent breakthrough was made by researchers at the Sandia National Laboratories in the US who announced in December 2009 that they had successfully tested a machine that can transform CO_2 into fuel using solar energy.

Wind energy

Solar radiation heats the atmosphere above land and sea unequally and thus creates zones of low and high pressure, which move air masses. Between the poles and the equator, the sun thus heats the earth in very different strengths. The differences in temperature that result stimulate the differences in the density of the air masses in the high-pressure zones towards the low-pressure zones. This constitutes the general phenomenon of winds moving across the surface of the planet.

Wind is therefore a mass of moving air that transforms the thermal energy that it receives from solar radiation into kinetic energy. Wind has two essential parameters, the first, which is determinant for the quantity of energy that it can supply, is its speed, and the second is the direction of its movement. Speed and the direction of winds can be strongly influenced by local conditions, particularly the geographical relief and nearby obstacles.

A wind turbine transforms the kinetic energy of the wind into mechanical energy. Either this energy is used directly as with windmills for pumping water or it is transformed into electricity though a generator. In the latter case it is called an aerogenerator. Two different usages are possible. The main one is coupling the aerogenerator to the network, the field in which there has been the most research and experiments made because it is the most effective. The second one is using the installation as a wind electricity generator. This is above all used in isolated regions.

According to Lester R. Brown, to cover 40 per cent of the demand for energy in the world, it would be necessary to install 1.5 million wind turbines of 2 MWs. With 65 million cars produced each year, he believes this is not an impossible task. Evidently it is a political choice, but there are limits. To cover all the energy production of France, would require the territory of two whole *départements* while Germany, with its 20,000 windmills, has reached the limit of the possible with the present constraints.

The installation of wind turbines must necessarily respect certain conditions : they must be situated on a plateau or a gently sloping hill (the speed of the wind increases with the height), on a surface that is clear and regular, and sufficiently far (at least 100 metres) from obstacles both natural (trees, changes of level) and artificial (houses, walls, posts, etc.). These obstacles create wind, and thus turbulence that considerably upsets the regular rotation of the wind turbine's blades and, after a short period, they can destroy the machine. Finally, the wind turbines must be oriented towards the dominant winds and hence the need to measure, besides the speed of the wind, its direction.

Types of wind turbine

There are two families of wind turbine at the present time, those with a vertical axis and those with a horizontal one. Wind turbines with a vertical axis do not need wind orientation systems, but their design is usually rather complicated.

Wind turbines with a horizontal axis (a helix) are more flexibly designed, have a high yield and are consequently more widespread. Their common characteristics are that they are mounted at the top of a pylon and equipped with a system of wind orientation. According to the number of blades contained by the helix they can be divided into two groups: ones with a 'multiple' slow rotation and the aero-generators with rapid rotation.

Those with the 'multiple' slow rotation, which have been used for a long time in rural areas, in France e.g., serve exclusively for pumping water. Those with rapid rotation, with two or three blades, are currently preferred and they are usually utilized for the production of electricity – and hence their usual name, 'aerogenerators'. When they have a certain power capacity, they usually have a helix with variable paces. In this case the position of the blades vis à vis the wind can be modified, making it possible to maintain a high yield whatever the speed of the wind and the rotation speed of the wind turbine.

The energy yield of a wind turbine varies in function of the speed of the wind per cube. At the present time they can support winds of 200 kilometres an hour – and yet only two years ago, they could not support winds over 90 kilometres an hour. There is thus considerable progress. Nevertheless, wind energy requires supporting energy for the less windy days and a storage arrangement (battery) to accumulate the energy produced.

Apart from the numerous advantages that it shares with other sources of renewable energy, the exploitation of wind energy has certain characteristics: it is adjustable and can very well be adapted to the capital available, as well as to the energy requirements. Thus there are no superfluous investments. Operating costs are quite limited given its high level of reliability and the relative simplicity of its technology. The cost prices will probably diminish in the coming years, following economies of scale in their manufacture.

Technically well designed, wind turbines are profitable in regions with a lot of wind. In addition, the period of high productivity, which is often during the winter when the winds are strong, corresponds to the period in the year when the demand for energy is greatest. Furthermore, they can be installed off-shore.

Nevertheless, there are some disadvantages. Although the cost of installation is clearly decreasing, it still remains high and hence out of reach for developing countries. The installation must be precise and rigorous, as it involves the situating and assembling of a tower of 10–30 metres high. Its exposure to regular wind (strength and direction) is very important. An irregularity or an absence of wind for several days can pose problems and there must be a good storage capacity (plus a large number of batteries). Finally, the use of many construction materials (metal, cement, etc.) makes it expensive and also ecologically costly.

The effects of wind farms on the local environments must be studied carefully at the planning stage. In general, any negative impact can be overcome by technical and aesthetic solutions that do not affect the viability of the project. At a regional level, studies that look at both wind potential and environmental values make it possible to identify the best regions for installing wind farms and those where their development would be harmful for the environment. There are thus appropriate solutions in this field, but they are limited in terms of the world demand for energy.[7]

Hydrogen energy

Hydrogen batteries are proposed as a way of replacing petrol in vehicles. In fact, hydrogen is the most abundant element in the universe. Of course it is necessary to produce dihydrogen (H_2), as opposed to oil, which only has to be extracted from the subsoil. Hydrogen gas is used in making a battery that can be used as a source of electric energy.

While the classic batteries transform chemical energy direct into electrical energy in a discontinuous operation, the combustible battery aims at transforming chemical energy into electrical energy, but continuously and with cheaper combustibles. The first fuel cell batteries were developed to feed the Gemini space capsules, of which the first unmanned probe took place on 8 April 1964, and the first manned one on 23 March 1965. Since then the aeronautic industry and space industry have remained the main users of this type of generator.

Composition and functioning of hydrogen batteries

A hydrogen battery is a combustible battery using dihydrogen and dioxygen, with the simultaneous production of electric current, water and heat, according to a chemical reaction in the functioning of the battery. In its natural state this will explode but it can be controlled and used to produce exploitable energy. Each electro-chemical elementary cell supplies about 1.2 volts of tension. The intensity depends on the quantity of matter (of dihydrogen and dioxygen) that has been introduced. But these electro-chemical cells are not enough to feed the engine of a car. They are thus put together in series or in parallel so as to increase the power supply. Bipolar plates make it possible to bring together the elementary cells; their function is to distribute gas (H_2 and O_2), to recover the water, chill the core of the battery (the membrane) and collect the electric current. The stacking up of elementary electro-chemical cells is usually called a 'module' or a 'stack'.

$$2 H_2 \text{ (gas)} + O_2 \text{ (gas)} \ 2 H_2O \text{ (liquid)} + heat + electricity$$

The hydrogen battery or fuel cell battery has several advantages. First, even when produced from fossil fuels, it is an energy that is cleaner than oil. It emits only water and thus helps to fight pollution. Dihydrogen also possesses an energy power much higher than oil (120 megajoules per kilogram (MJ/kg) for dihydrogen as against 45 MJ/kg for oil).

But there are also disadvantages. Dihydrogen is not yet a perfect fuel: there are numerous problems still unresolved. For, in fact, it is produced 95 per cent from fossil fuel. Moreover, it is much more difficult to store dihydrogen than oil. To do so it is necessary either to compress it under high pressure, which means using voluminous heavy tanks, or to liquefy it at a low temperature (which means problems of thermal isolation). Also, certain of its physical properties (inflammability in the air) make storage complicated.

While fuel cell batteries are not yet economically competitive, mainly because of the cost of the electrode-membrane-electrode (EME), they are advantageous over other technologies in that they give a higher energy yield (usually over 50 per cent as against plus or minus 30 per cent for thermal motors), cause far fewer polluting discharges and considerably reduce noise levels. The fuel cell battery is a well-designed technology: the batteries already exist and current research is only concerned with improving their characteristics.

In addition, the hydrogen battery only emitting water enables it to be used on a massive scale to help resolve the problems of greenhouse gases. Nevertheless, it is necessary to develop the means for producing dihydrogen and setting up a distribution infrastructure, which will be necessary, however costly. The hydrogen battery can thus enable a car to become independent of oil.

The scientific threshold has been crossed and what remains to be done is its application, with all the economic and political implications. The electricity thus produced by this technology can be used to run an automobile. Thanks to a partnership between Total and BMW, a research group has been established to conceive a hybrid car that functions with hydrogen and petrol. Such a car, when perfected, should have an autonomy of 200 kilometres and beyond that petrol will automatically take over. Some countries are well advanced in this field. Already in 2007 Germany disposed of equipped service stations in Munich, Hamburg and Berlin. In the German capital, in the same year, the total number was to pass from 16 cars to several hundred, thanks to a public/private partnership.[8] For a carmaker like BMW, the ecological solution will come from mixed formulas, associating economies of energy, agrofuels of the second generation, a hybrid motor, hydrogen and electricity. According to the US Academy of Sciences, the car industry could put 2 million vehicles running on hydrogen into circulation between now and 2020.[9] Hydrogen has a future, but it will probably be several decades before it is widely applied.

Energy from waste

Waste of agricultural, domestic and industrial origin can serve as raw material for the production of electric current, as well as of methane gas which, after treatment, can have several uses. We should remember that the incineration of domestic waste makes it possible to recover only a part of the energy that has been used in its production, so that it is wiser to limit production at the source, to avoid waste of energy.

It is possible to divide waste into two categories: industrial waste, the residues of hydrocarbons, tar, used solvents and other sludge produced by industry and which can be transformed in special incineration centres into heat or electricity; and agricultural and agro-industrial waste. The latter includes, for example, the straw of the most widely-cultivated cereals in the world, wheat, maize and rice. It is possible to get 2–6 tonnes of straw from 1 hectare of cultivation. Agro-industrial waste comes mostly from sugar refineries and oil mills.

Waste becomes decomposed through heat (pyrolysis) which produces gas fuel. This is burnt at a temperature of 800–900 °C in a post-combustion chamber. From 5 to 7 tonnes of waste are necessary to produce the equivalent of 1 tonne of fuel. The energy of the incineration oven is captured to heat water, produce vapour and generate electricity or else it is exploited in co-generation (heat and electricity).

There are two main ways for transforming waste into energy: incineration, where the waste is burnt, producing heat, electricity or both (co-generation) and methanization (anaerobic fermentation) or the transformation of waste of organic origin and carbonic gas (biogas).

The methanization of organic matter

Methanization is an anaerobic digestion, or methanic fermentation, which transforms complex organic matter into compost, methane and carbonic gas by a complex microbial eco-system that functions in the absence of oxygen. It makes it possible to eliminate organic pollution while consuming little energy, producing only a small amount of sludge and generating renewable energy, i.e. biogas. Methane representing 55–85 per cent of the volume of the biogas produced can be used as a source of energy. One cubic metre of methane (8.750 kilogram calories) is the equivalent of 1 litre of fuel oil.

Methane gas (biogas) can be used like natural gas after treatment. It supplies industrial fuel for producing electricity and heat, petrol for cars or it can be injected into municipal gas systems. There are, however, considerable environmental disadvantages in the incineration of waste: atmospheric pollution (dust, acid gas, dioxin, heavy metals, etc.) and greenhouse gases. The residues from the cleaning of the chimneys are toxic. This process can also take place in a digester, to decontaminate the refuse changed into organic materials while producing energy in the form of methane.

There are some studies on the exploitation and utilization of methane gas from collieries, usually known as firedamp. Prospecting has been carried out in the coal basins of Lorraine in France and also in Wallonia (in the mining sites of Charleroi and Borinage), which is considered the cradle of European mining.[10] These experiments have real potential but are evidently limited to existing reserves.

Geothermal energy

From the Greek *geo* (earth) and *thermie* (heat), geothermy is the science that studies the internal thermal phenomena of the terrestrial globe and the technology that aims at exploiting it. By extension, geothermy also applies to the geothermal energy that comes out of the earth and is convertible into heat.[11] This comes from the disintegration of radioactive elements in rocks and in the terrestrial core that generate a flow of heat towards the surface. The greater the depth, the greater the heat, increasing on average by 3° C every 100 metres. But this geothermal gradient can be much higher in certain geological configurations. In seismic or volcanic zones the 'thermal gradient' can be ten times greater and even reach 100 °C in certain places.

The three types of geothermal energy take the heat contained in the soil which varies according to the depth level. This energy has been exploited in heating and hot water networks for thousands of years in China, in Ancient Rome and in the Mediterranean. The increase in the price of energy and the need to emit fewer greenhouse gases make it more attractive these days. In several countries it is being used, as in Nicaragua, from the Momotombo volcano and on the island of Reunion.

The advantages of this source of energy is that it is free, renewable and that its exploitation is not expensive. The installations that use geothermal energy do not pollute the atmosphere. Co-generation, that is the production of electricity at the same time as heat can still further increase interest in it. However, it is an energy that is difficult to transport and it must be used in situ. Very substantial investments are often needed to pump hot water.

The main use is for heating housing and other buildings. But geothermy can also heat greenhouses, serve pisciculture,[12] livestock raising, the drying of agricultural products, and be used to prevent roads from freezing (pipes of hot water under the asphalt) and for air conditioning and refrigeration.

Geothermal energy has a weak impact on the environment. It emits little CO_2. However, it is necessary to take into account the gases that are contained in the water, but that can be removed, methane, sulphuric hydrogen, CO_2, etc. Also, the geothermal water must not be released into nature as it contains salts and heavy metals. These risks are not great if the water is reinserted into the subsoil. Once again, a technically interesting solution remains limited in its application.

Three types of geothermal energy

Geothermal energy that is not very deep or relatively cool corresponds to temperatures between 30 and 100 °C. These sources are to be found at depths between 1,000 and 2,500 metres in formations that are rocky and permeable, full of water and usually situated in very large sedimentary basins. Investment in geothermy is around 400–600 euros for each kilowatt installed, but the cost of operation is very low: from 0.05 to 0.1 centimes (of euros) per thermal kilowatt hour.

Deep geothermal energy at a high or 'average' temperature corresponds to the hot water sources under pressure, the temperature of which is between 90 and 180 °C. They are to be found at some hundreds of metres and several kilometres deep. Machinery can now be installed that can produce quantities of electricity from a few kilowatts to some megawatts, representing an investment between 1,000 and 4,000 euros per kilowatt installed which will last for 30–50 years. They are to be found in sedimentary basins, which are the most suitable zones for low-energy geothermy, but at deeper levels, from 2,000 to 4,000 metres in many very localized zones. In such cases it is usually hot water that comes up from deep down through geological faults.

Very deep geothermal energy at very high temperature or 'high energy' exploits deposits of dry or humid steam (a mixture of water plus steam), situated between 1,500 and 3,000 metres down in volcanic zones or at the frontiers of tectonic plates where geothermy is particularly intense, and its temperature is between about 200 and 350 °C. The cost of a kilowatt hour varies according to the method used but is between 4 and 7 centimes (of euros).

Agrofuels from Photosynthesis

Agrofuels certainly have a role to play among the future solutions but it is much smaller than was previously supposed. In fact, as we have seen, the frantic rush to develop them has provoked growing opposition and their benefits are increasingly contested. The main stumbling block is the energy balance sheet of their production, that is, the difference between the quantity of energy necessary for a complete production cycle and the quantity of energy reproduced by agrofuels during their utilization as fuels. Various studies have researched the question, but with different results. This difference has depended on the type of agrofuel being studied (such as ethanol, vegetable oils), the origin of the plant (wheat, maize), the method of production (whether the transformation site is near the field of production) and the place of cultivation (Europe, Brazil, Indonesia).

According to a study published in *Science* in August 2007 and carried out by Renton Righelato of the World Land Trust and Dominick Spracklen of the University of Leeds, it would be better, from the greenhouse gas viewpoint, to protect and restore the forests and prairies than to use this land to produce agrofuels. Furthermore, on the basis of a simulation exercise carried out over a period of 30 years, it has been calculated that the replacement of forests by cultivation to supply cars with flex fuel would emit up to nine times more CO_2 during the same period. Serious warnings concerning agrofuels also come from Paul J. Crutzen (1995 Nobel Chemistry Prize winner) of the German Max Planck Institute. The results of this study, co-signed by an international team of researchers and published in the *Atmospheric Chemistry and Physics Journal* of September 2007, were that the production of a litre of agrofuel can contribute up to twice as much greenhouse gas than the combustion of the same quantity of fossil fuel. The team of researchers particularly stressed that the emissions of azote protoxyde (N_2O), the greenhouse gas discharged by intensive agriculture is 296 times more powerful than CO_2. In fact, a non-negligible part of the azote fertilizers used to increase yields dissolves into N_2O.

Another warning comes from the Organization for Economic Cooperation and Development (OECD) which expressed its concern in September 2007 at a round table on sustainable development. The OECD calculates that, of the current technologies, only agrofuels produced from cane sugar, cellulose, animal fat and used kitchen oil can make a real reduction in greenhouse gas, as compared with petrol and diesel. The other techniques of production can theoretically bring about a reduction of 40 per cent of greenhouse gas, but if acidification of the soils, use of fertilizers, loss of biodiversity and toxicity of pesticides are taken into account, the negative global impact of ethanol and agrodiesel is much greater than that of petrol and diesel – not to mention the social effects about which we have spoken at length.

From all this it can be concluded that the alternatives in no way make it possible to replace fossil fuels in the short and medium term which will remain dominant in the years to come, constituting at least 80 per cent of the total. In other words, they will continue to have negative climate effects, even if the new technologies manage to attenuate them. In spite of the progress made, their gradual disappearance cannot be compensated in the medium term by all the alternative energies, which does indeed give ground for concern. The spread of agrofuels along the lines envisaged can only aggravate ecological and social problems and it is imperative to halt them. There remains only one valid long-term solution, which is to change our ways of consuming energy, but that is contradictory to contemporary economic logic. Hence the need to pose the problem of an alternative development model, faced as we are with what can be called a veritable crisis of civilization.

POST-CAPITALIST LOGIC OF THE ECONOMY AND A NEW DEVELOPMENT MODEL

It is not only a question of identifying and applying short-term solutions to resolve the energy issue. There has to be a transition that makes it possible both to transform the development model based on exploitation without counting on energy sources, as well as finding and developing new resources and new technologies. One cannot go ahead without the other and it would be illusory to believe that scientific and technical improvements can resolve the problem within the framework of capitalist logic. By small or large steps, the way forward has to be based on a post-capitalist logic. The rational way of doing it seems to be the most reasonable one, rather than waiting for a global crisis to force through a solution. Hence the necessity of a general framework that includes a solution to energy problems.

Such a project would be based on four main principles corresponding to the fundamental elements of human existence.

For Sustainable Use of Natural Resources

In a relatively short time, we have seen a veritable explosion in the collective consciousness of the fact that the indiscriminate use of natural resources is endangering the continuity of life itself, physically and biologically. Not only is humanity confronted by the exhaustion of certain natural riches, it must equally tackle the destruction of elements that are essential to life, such as soil, air, atmosphere and climate. The more industrially developed societies consume three to four times the theoretical possibilities of the ecological renewal of the planet. Re-establishing a balance in the use of natural resources has thus become a question of survival. Moreover, the non-renewable sources, particularly in the field of energy, cannot be left to the commodity logic, and hence add to the accumulation of capital. Rather, it must be possible to administer them collectively so as to contribute rationally to the well-being of humanity.

All this means that a new philosophy of the relationship between human beings and nature is required. We should pass from the notion of exploitation to that of symbiosis. This is indeed contrary to the idea of endless progress, nature being inexhaustible, according to the legacy of the Enlightenment. It also means rediscovering certain values in the thinking of the pre-capitalist societies, particularly the fundamental unity between humanity and the natural world and solidarity as the basis of social construction. Obviously, the genuine progress of analytical thinking must be taken into account, replacing the causalities and operational mechanisms of nature and of society in the physical and social fields, thus going beyond mythical thinking and identifying the symbol with reality. Such a perspective also means distancing ourselves from the socialism of the twentieth century, which was strongly influenced by scientism and by a linear vision of progress.

The end of the twentieth century has been marked by a criticism of modernity, which has entered into the social sciences through the influence of new orientations in the natural sciences. This is particularly so in the introduction of concepts of complexity and incertitude, as in the work of Ilya Prigogine, showing the role of hazard and unpredictability in both physical and biological sciences, but also in the social sciences, as in the writings of Edgar Morin. The latter has developed a critical position, avoiding the post-modernist radical who denies the existence of systems and structures in favour of immediate history constructed by individuals and who opposes 'small narratives' as against 'great narratives' – in other words, explanatory theories. Morin does of course acknowledge the reality of the unpredictable and the uncertain in social sciences, but he affirms the existence of a fundamental paradigm that is to be found in the physical, biological and anthropological worlds – i.e. the constant passing from disorder to self-reorganization, or the continuity of life.

What is at stake today, according to Edgar Morin, is the very possibility of reorganization. Human activity produces irreversible effects that have catastrophic consequences on the natural world and on human groups themselves. This French sociologist and philosopher has even reached very pessimistic conclusions, believing that it is probably already too late to change the course of things. It is not, however, necessary to go to such an extreme to become aware of the need for a radical reaction. This evidently poses an ethical problem: the need to ensure the reorganization process in different fields. It is indeed a question of life itself, as Enrique Dussel illustrated so well in his work *L'éthique de la Libération*.[13]

This was also understood by a number of social actors when, in 2004 in Mexico, they started up a network of intellectuals and artists for the defence of humanity. The question of energy is of course directly linked to these problems. To the extent that its production and utilization contributes to the aggression against the reproduction of life, it cannot escape the fundamental question of relationships with nature.

Giving Priority to Use Value rather than Exchange Value

These concepts were elaborated by Karl Marx and have entered into common use. The use value is that which concerns products or services for use by human beings, and the exchange value is that which these elements acquire when they enter into the market. The characteristic of capitalism is to favour the exchange value as an engine for economic development. It is logical, because only exchange value makes it possible to make a profit and thus give rise to an accumulation process.

As a result, the market becomes a natural phenomenon, which is no longer considered a social relationship. The priority of the market becomes a dogma, from which everything else follows automatically: it imposes its logic on all human collective relationships and all sectors of activity. Its laws apply even to sectors like health, education, sport and culture. Such logic excludes parameters other than those of economic exchange, particularly of the qualitative kind, such as the quality of life or what are called externalities – everything that precedes or follows market logic, enabling all the costs not to be taken into account, including the production of energy. Giving priority to use value thus means favouring human beings over capital. Such a priority brings about a number of consequences, of which we can cite just a few.

If the use value predominates, the life of products would be prolonged. This, according to Wim Dierckxsens, a Dutch economist working in Costa Rica, has numerous advantages. In fact, to accelerate the circulation of capital and contribute to its accumulation, the life of products has been reduced. Prolonging their life would mean using fewer raw materials and less energy, producing less waste and therefore protecting the natural environment better. It would also reduce the influence of finance capital.

The same logic would make it possible to accept different prices for the same industrial or agricultural products, according to the different regions in the world. At present, the law of the market demands that world prices be aligned at the lowest level particularly in agriculture, which means the prices of the regions that have adopted capitalist productivist agriculture (often subsidized and practising dumping). In contrast, the arguments for use value can justify different prices, which goes against the dogma of the market. Why should rice have the same price in the US and Sri Lanka, if in the latter country rice forms part of its history and culture and if its production is a requirement for food sovereignty? Such considerations are discounted in market logic, but are taken into account in use value reasoning.

It would therefore be possible to relocalize production and avoid much cost of transportation, which harm the environment and cause, in many parts of the world, a congestion in communication routes and even paralysis of the roads. Applying use value would also make it possible to favour peasant agriculture, which creates jobs. In the service sector, education would be redefined primarily in function of people and not of the market and the production of medicines would have to take into account the number of diseases existing in the whole world and not in function of profits from sales.

To prioritize use value thus means to centre things around human life. It would be impossible to ignore the destiny of the 20–30 per cent of the world's population living in destitution. It would also be possible not to allow the rest of the population to become vulnerable, outside the privileged ones, because it would be human needs that become the motor of the economy. Also it would inevitably signify the establishment of mechanisms for the redistribution of wealth and generalize people's security of existence. As for energy, it becomes a use value aimed, in accomplishing the first principle of respect for nature, at satisfying the real needs of human beings and not to serve the accumulation of capital.

Such a perspective will obviously require a new economic philosophy. It would no longer be possible to define it simply as an activity producing added value, but rather to consider that its function, as we have already said, is to produce the material basis necessary for the physical, cultural and spiritual life of all human beings all over the world. Finally, this leads to an ethic of life, that is, the need to ensure its vital basis for everyone.

Generalizing Democracy

Generalizing democracy refers to all human relationships. Obviously politics is the first field of application. Representative democracy is glaringly deficient and today it has reached a point of non-credibility in many regions of the world, which is shown by the high levels of abstentions in those countries where the vote is not obligatory. It is therefore necessary to complete the representation by other mechanisms, which are now called 'participatory'. Even if this term has become vague and ambiguous because it has been misused, the concept is still fundamental. It is a question of expanding the space of citizen responsibility.

There are numerous possible formulas, from the famous participatory budget, as it was initiated in Porto Alegre, Brazil, to the regular control by voters of their representatives through the presentation of accounts and even referendums. It also involves the elimination of lobbies, of the predomination of money in presenting candidates for public positions and, obviously, transparency in the composition of the electoral lists and in the operational mechanisms of political parties.

But it is not only in the political field that democracy has to be generalized. It should prevail in all the places where social relationships are built, from gender relationships, which should be based on equality, to the very functioning of social movements and finally, production relationships. There is nothing more anti-democratic than the capitalist organization of the economy and that is patent, from the business enterprise to the intermational financial organizations. The same logic dominates everywhere, that of the prevalence of the exchange value and thus the almost exclusive decision-making power of capital. A democratic process could no longer link the economic decision to the private ownership of the means of production. There are obviously many ways of envisaging the democratic economic process and they are not necessarily linked to the nationalization of all sectors of the economy. There are various ways of ensuring a democratic way of operating, including cooperatives, producer associations, ownership by a local community, etc. Hence the importance of redefining the state and its functions. Once again, energy is at the heart of the problem, because its collective control at the different levels of power is the only guarantee that it will be rationally used.

A new philosophical approach is therefore necessary. The typical characteristic of a democracy is the dialectic between creativity and organization. It excludes or strongly downgrades avant-gardism. It considers all human rights as a possibility for participation, without forgetting that the first human right is the right to live. Such a philosophy also recreates the centrality of the relationship between the individual and the collective. As for the ethical dimension of this third principle, it concerns respect for democracy within each system of social relationships, whether it be political parties, businesses, social movements and all cultural institutions, not forgetting gender relationships.

Multiculturalism

The fourth principle concerns multiculturalism and intercultural-ism in which all cultures, knowledge, philosophies and religions participate in the construction of 'a new possible world'. This means coming up against the cultural hegemony of the West, not only at the economic level with the imposition of the capitalist model, but also at the level of values. Interculturalism obviously cannot be imagined without integrating the three other principles just described, their unity being indispensable. There would be no question of accepting a philosophy that contains racist principles or a religion that preaches the inferiority of women. The way in which energy is represented in many traditions and the ethical development which accompanies it constitute an enormous heritage capable of transplanting the issue of its role in human development. This is what we tried to show in a study carried out for the United Nations Educational, Scientific and Cultural Organization (UNESCO)[14] which was concerned chiefly with respect for nature, moderation in consumption and the ethics of solidarity.

In the framework of these four great principles it is therefore a question of respect for cosmovisions, or ways of interpreting reality, allowing all the cultural wealth of humanity to contribute to the common good, rather than being reduced to the isolation of ghettoes. Such a position evidently requires a philosophy of inter-culturalism as a cultural dynamic, an open conception of culture and its possibilities of transformation. This also needs a secular conception of the state as guarantee in intercultural participation. Finally, the ethic in this field would mean mutual respect, dialogue and collaboration in many social and cultural initiatives.

Building up such a post-capitalist model, which some call the socialism of the twenty-first century, is an initiative that benefits from past experience as well as new sensitivities that have been expressed through the social movements of the new generation and that stress values and the qualitative aspects of life, as well as democracy as a means and not only as an end. At the same time it will be a continuing construction because there has been an accumulation of thinking and doing that are rich in lessons. It is with this perspective that the project can arouse hope and enthusiasm, going well beyond the petty calculations of partisan actions. It is thus worth the trouble to pursue a struggle that ends in the construction of alternatives and to develop the critical thought that this requires.

The solution to the double crises of energy and climate is to be found in a global vision of the changing of civilization and not only in a collection of technical solutions. It is the only way that humanity can adopt a path that will permit its survival. Combining this radical change of society with immediate measures permitting the economy of energy with using new sources of energy that respect nature and social relations: these form the basis of the policies to be followed.

Glossary

acidification   phenomenon caused by emissions of azote oxide, sulphur oxide and ammonia by reaction with water when added to the preparation

agrodiesel   diesel obtained from vegetable oils: colza, oil palm, sunflower, soya, jatropha, etc.

agrofuel   a fuel that can be used in combustion engines that comes from the transformation of vegetable origin, either from oleaginous crops from which pure oil can be obtained (by crushing the grains of colza or sunflower, for example) and directly usable in diesel, or the transformation of vegetable oil or the fermentation of vegetable sugar producing alcohol (ethanol)

albedo   surface reflectivity of the sun's radiation

antioxidation   a chemical compound that can prevent the oxidation of organic matter

aromatic component   a fragrant component of which benzine forms the core of its structure

autotrophe   an organism using light energy to synthesize sugars and proteins from inorganic substances

azote protoxyde (N_2O)   a powerful gas with a greenhouse effect, having a global heating power over 100 years, which is 310 times higher than an equivalent mass of CO_2. The main cause of N_2O are natural phenomena of nitrification/denitrification in cultivated soil, particularly through the use of mineral azote fertilizers and the management of animal excrement

B100   pure agrodiesel

B20   a fuel composed 20 per cent of agrodiesel and 80 per cent of fuel oil

bagasse   fibrous residue left after the extraction of sugar-bearing juice from sugar cane which can be used in a boiler to produce electric current

barrel of crude   a measuring unit of petroleum

biobutanol   a mixture of agrofuel and petroleum

biodiesel   *see* agrodiesel

bio-economic   economic activity based on the energy obtained from renewable energies or agricultural activities

bio-ethanol   fuel produced from vegetable matter such as sugar cane, maize, wheat, etc.

biofuel   *see* agrofuel

biogas   gas produced from fermentation of organic matter in the absence of oxygen, through bacterial action, and which can feed fuel cells. The operation can produce 90 per cent gas and 10 per cent of CO_2 and water

biomass   all living and recently dead biological matter (branches of trees and stalks, vegetable waste, liquid manure, animal droppings, household and food industry waste) which makes gas by fermentation because of action by micro-organisms. It can produce heat or energy

bioplastic   plastic made from renewable matter or agricultural activity

bioprospection using natural enzymes (for example wood ants) to free the sugar from wood and produce ethanol

bio-refinery a factory producing fuel based on vegetable matter

bitumen a mixture of hydrogen carbide which can be solid or liquid. Artificial bitumen is obtained in the distillation or oxidation of petroleum

calorific power the maximum energy that can theoretically be obtained from a kilogram of matter

carbu-modulable vehicle a vehicle that can function with various kinds of fuels

catalytic hydrogenation a chemical reaction consisting of the addition of pairs of hydrogen atoms to a molecule. It requires the use of metallic catalysers like platinum (Pt), nickel (Ni) and palladium (Pd)

cation an ion having lost one or several electrons

cellulose matter constituted by vegetable cellular membrane

CH_4 the methane gas contained in the atmosphere that contributes to the greenhouse effect

CO_2 carbon dioxide, a gas of natural origin or the result of combustion of fossil fuels and of biomass, as well as of changes in soil use and of other industrial processes. It is the main greenhouse effect of human activity that influences the net balance of sun rays on the surface of the earth. It is also the leading gas by which all the other gases with a greenhouse effect are measured from 1, and which therefore potentially affect global warming

coal vapour the equivalent in coal of heat or electricity

co-generation the simultaneous production of heat and electricity. It is called micro co-management when it produces both for domestic purposes and for a network

colza an oleaginous plant the grains of which can be used for producing agrodiesel (1,500 litres per hectare)

dead zone an area in the ocean that lacks oxygen brought about by the draining of nitrates into the water courses or from azote fertilizers used particularly for maize and sugar cane. These nitrates cause the development of algae which sink to the bottom of the sea when they die and while decomposing use almost all the oxygen in the water and killing aquatic life

dendroenergy energy from ligneous (woody) fuels. Wood and charcoal are the fuels most utilized in the food processing industries for cooking, mixing, smoking, drying and the production of electricity

dihydrogen a hydrogen molecule with its two atoms (H_2)

dioxygen an oxygen molecule with its two atoms (O_2)

E85 fuel composed of a mixture of 85 per cent ethanol and 15 per cent of leadless petrol

edaphic pertaining to the soil

electrical automobiles an automobile whose main energy is supplied by a battery that stocks electricity

electrolysis the production of a chemical reaction due to an electric current

endosulfan the active substance of a phytosanitary product (phytopharmaceutical or pesticide) which is an insecticide belonging to the organochlorine chemical family

ester a chemical body resulting from the reaction of an acid (glycerides, for example) and an alcohol, with the elimination of water

ethanol   a fuel produced from alcohol plants, such as maize, sugar cane, beetroot, wheat, etc.

ethyl tertio butyl ether (ETBE)   is commonly used as an oxygenate petrol additive in the production of petrol from oil. It can be incorporated up to 15 per cent in volume in the petrol

euphorbe is a perennial plant that contains a milky juice which becomes dark when exposed to air and is capable of producing oil (like jatropha)

eutrophization of water   a term that originally identified its richness in nutritional elements, with no negative connotations.   As from 1970 the term has been used to qualify the degradation of large stretches of water through an excess of nutrients

exchange membrane   used for exchanging ions and protons in batteries

floculant   a polymer (a large molecule composed of repeating structural units typically connected by covalent chemical bonds) that imprisons accumulated colloidal matter thus forming big flakes that settle as sedimentation

fossil fuel   fuel that originates from coal, gas or oil

fuel cell   a reserve of energy that can use hydrogen as a fuel in order to produce electric current

furfural   a natural bactericide, fungicide and insecticide to be found especially in the African palm, of which the leaves contain 17 per cent

gasification   the process of transforming into gas, for example coal into gas in the mine

glycerides   lipids formed by the mixture (ester) of one, two or three fatty acids with alcohol – glycerol (or glycerine)

green gold   agrofuel

greenhouse gases   all the gases (carbon dioxide, methane, steam, azote protoxyde, etc.) contained in the atmosphere which, by absorbing and re-transmitting infrared rays emitted by the earth, contribute to the rise in temperature

hybrid motor   an engine that recovers the thermal energy lost by the vehicle through electricity stored in the batteries during descent or braking

hydroelectricity   electricity produced from running water

hydrolysis   decomposition by water

ITER   a thermonuclear fusion experimental reactor installed at Cadarache (Bouche de Rhone) in France

Jatropha curcas   a plant of the *Euphorbiaceae* family found in arid zones, the grain of which can be used to produce oil. Its oil cakes can serve as organic fertilizer

joule   is the derived unit of energy in the International System of Units. It is energy exerted by a force of 1 newton acting to move an object through a distance of 1 metre

kerosene   fuel used by aeroplanes

Kyoto (Protocol)   the result of the UN climate conference organized in 1997 in Kyoto (Japan). It engages its signatories to diminish the production of greenhouse gases until 2012. The US did not sign. China and India and some countries from the South signed, but were not obliged to make reductions. The conference was followed by another in Bali in 2007

La Via Campesina   an international movement composed of peasant, small and medium farmer organizations, agricultural labourers and women, as well as indigenous communities in Asia, Africa, America and Europe. It is an autonomous movement, pluralist and independent of all political, economic or other interests. It also consists of national organizations. The movement is organized into eight regions: Europe, North-East Asia, South-East Asia, South Asia, North America, the Caribbean, Central America, South America and Africa

lignocellulose   the basic component of wood, straw and grass

lithium batteries   disposable (primary) batteries that have lithium metal or lithium compounds as an anode. Their cells are 1 mm in diameter and are also called micro-batteries

LPG   liquefied petroleum gas which does not emit particles and less carbon dioxide than fossil fuels

marine current turbine   a producer of electricity from sea and river tides. Its flippers can turn either vertically or horizontally. It was installed for the first time in 2006 off the coast of Ireland

methanol   alcohol obtained by fermenting vegetable sugar, popularly known as wood alcohol

methyl ester   produced by the trans-esterification of the triglycerides of vegetable oils (palm, colza, sunflower, etc.). The vegetable oil methyl ester is a mixture of colza or sunflower oil with methyl alcohol obtained by fermentation, either using sugar (sugar cane or beetroot) or starch (wheat) and which, by hydrolysis, produces glucose

monoculture   cultivation of just one crop for many years

Montreal (conference)   meeting held in 2005 to revise the Kyoto agreement which prolonged the Protocol beyond 2012

nanotechnology   the utilization of the infinitely small

neem   a sacred tree originating in the south of the Himalayas in India, which is endowed with numerous virtues. It can adapt to poor soil, tolerates high temperatures and survives low rainfall. It grows rapidly, to reach a height of 20 metres and lives for 200 years. The kernel extracted from the nut is made into oil which is used as a fertilizer, pesticide and insecticide. This oil is now being used to produce agrofuel

N-P-K   a mineral fertilizer in which azote, phosphorus and potassium are the active elements

off-shore oil   petroleum extracted or exploited in the sea

oil   a liquid product from colza, sunflower, palm tree, etc. which is used, apart from food, pharmaceutical and cosmetic purposes, for diesel engines

oil protein plants   produce a vegetable oil used for food

oleaginous plant   a plant of which the fruit serves to produce oil

organic agriculture   agriculture that does not use chemical fertilizers

photosynthesis   a synthesis of carbon hydrates by vegetable chlorophyl through the light of the sun

photovoltaic cell   a panel to capture solar energy for transforming it into electrical energy or heat. Made from crystalline silicon and, more recently, from solar polymer film

phytosanitary product is used to protect plants. It is an active substance or an associatioon of several chemical substances or micro-organisms from a binder or possibly a solvent with an additive

polyelectrolysis electrolysis of several chemical components (*see* cation for catonic polyelectrolysis)

polymerization the union of several molecules of a component to form a large molecule

pyrolysis the chemical decomposition of organic materials by heat in the absence of oxygen

Round-up a herbicide based on glyphosate produced by the Monsanto corporation which is claimed to be 100 per cent clean and biodegradable. The firm was condemned in France to a 15,000-euro fine for misleading publicity. Studies carried out in the US, Denmark, Colombia and Argentina confirm its toxic character

starch a glucide stored by vegetables, fruits and tubers in the form of granules

sugar substance stored in sugar cane which can be used to make ethanol

sugarbeet a sugary plant which is used to produce ethanol (1 hectare of sugarbeet can produce 7.000 litres of agroethanol)

synthetic fuel a liquid fuel obtained from coal, natural gas and biomass

taxon regroups all living organisms that have certain taxonomic characteristics or well-defined characteristics in common

TEP tonne equivalent of petroleum

thermal power station electricity plant powered by coal, gas or fuel

thermodynamics the study of the conversion of energy into work and heat and the relation to macroscopic variables like temperature and pressure

tidal waves energy produced from sea waves, like the Rance plant in France

tourteau the solid sub-product made from the crushing of cereals (for example peanuts, soya), rich in protein, used as animal feed, as well as fertilizer

trans-esterification the reaction of an ester and an alcohol to make another alcohol. It is a reversible reaction, catalysed by an acid or a base. To make the reaction more complete more R-OH is added, serving at the same time as a solvent

treethanol ethanol produced from wood cellulose

triglycerides (also called triacylglycerol or triacylglycerides) the glycerides into which the three hydroxyle of glycerol groupings are esterified by fatty acids. They are the components of vegetable oil and animal fats

tropospheric ozone a form of allotropic oxygen containing three atoms in the molecule (O_3), a blue, odorous gas that forms in the air when oxygen undergoes an electric charge

turbidity cloudy state of liquid

vegetable oil an oil produced by extraction from oleaginous plants

water hyacinth an aquatic plant found in rivers, canals and lakes in tropical regions

watt a unit of electric current (W) corresponding to the consumption of 1 joule per second (J/s)

Notes

PREFACE

1. Richard Greenwald, *Time*, 14 April 2006.

CHAPTER 1

1. Whereas food security limits its aims to finding the technical means to provide food for a country's inhabitants by any means, including buying food from abroad or accepting commodity dumping, food sovereignty is more demanding and it includes the following principles: food is a basic human right and it should be spelt out in national constitutions; a genuine agrarian reform giving landless people ownership and control of the land they work; protecting natural resources, especially land, water, seeds and livestock breeds; reorganizing food trade, food being first a source of nutrition and only secondarily a trade item; ending the globalization of hunger by multilateral institutions and speculative capital; social peace – food must not be used as a weapon; democratic control – smallholders must have direct input into formulating agricultural policies at all levels.
2. François Houtart and François Polet, *The Other Davos*, London: Zed Books, 2001.
3. *Le Monde*, 14/15 October 2007.

CHAPTER 2

1. Jean-Michael Bezat, *Le Monde*, 20 May 2008.
2. Ibid.
3. L'Atlas de l'Environnement, *Le Monde Diplomatique*, Paris, 2007, p. 7.
4. *Les Clés de la Planète: Agir pour la Terre*, Toulouse: Milan Presse, 2007, 3.
5. Oscar René Vargas, *Geopolitica en el Siglo XXI*, Managua: Certen, 2007, 73.
6. Quoted by Marc Roche, *Le Monde*, 11 August 2007.
7. Ibid.
8. *Le Soir*, 2 October 2007.
9. Ibid., p. 15.
10. *Le Soir*, 23/24 September 2006.
11. *Newsweek*, 25 April 2007.
12. Hervé Kempf, *Le Monde*, 15 March 2008.
13. *Le Monde*, 15 August 2006.
14. Patrick Moon, *El Pais*, 22 July 2006.
15. Super-phoenix is a fast breeder reactor, set up on the Rhone near the Swiss border. It halted electricity production in 1996 and was closed commercially the following year. It was abandoned because of a series of technical problems and high operating costs, while the widespread opposition to it on the part of ' nth French and Swiss anti-nuclear groups was also certainly a factor.

16. Gaëlle Dupont, *Le Monde*, 20 March 2008.
17. *Le Monde*, 12 December 2007.
18. Catherine Ferrieux, 'Un verdict sans appel' ('A verdict that cannot be appealed against'), *Sciences et Avenir*, March/April 2007, 6.
19. *Risal Information*, 12 March 2007.
20. Anthony Giddens, *The Politics of Climate Change*, Cambridge: Polity, 2009, 7–14.
21. Catherine Brahie et al., *Courrier International*, 20–6 September 2007.
22. Ray Weiss, *Le Soir*, 6 November 2008.
23. *El Pais*, 3 December 2007.
24. Vicente Barros, *El cambio climatico global*, Bogota: Desde Abajo, 2004, 46–65.
25. Michel Destrot et al., *Energie et Climat*, Paris: Plon, 2006, 14.
26. *Sciences et Avenir*, March/April 2007, 10.
27. 'L'ennemi climatique No. 1', *Sciences et Avenir*, No. 150, March/April 2007, 30.
28. IPCC, *Fourth Assessment Report (AR4)*, 2007.
29. UNDP, *Human Development Report, 2007–2008*, 11.
30. *Le Monde*, 12 June 2006.
31. Ibid.
32. *Sciences et Avenir*, March/April 2007, 8.
33. *Le Soir*, 11 March 2008.
34. Gaëlle Dupont, *Le Monde*, 5 December 2006.
35. *Wall Street Journal*, 10 October 2007.
36. Quoted by Stéphane Foucart, *Le Monde*, 29 September 2006.
37. *Le Soir*, 23 October 2007.
38. L'Atlas de l'Environnement, *Le Monde Diplomatique*, 2007, 33.
39. *Sciences et Avenir*, March/April 2007.
40. UNDP, *Human Development Report, 2007–2008*, 157–8.
41. *New Geographical Magazine*, 29 February 2007.
42. Scott Wallace, 'Les déchirures de l'Amazonie', *National Geographic Magazine*, February 2007.
43. *World Rainforest Bulletin*, No. 132, 30 July 2008.
44. The Monroe Doctrine (named after James Monroe, president at the time) was declared by the US in 1823, warning European powers not to colonize or interfere with states in the New World (particularly Latin America). In exchange the US pledged not to interfere with the internal concerns of European countries.
45. L'Atlas de l'Environnement, *Le Monde Diplomatique*, 2007, 36.
46. Pierre Friedlingstein, *Sciences et Avenir*, March/April 2007, 30.
47. United Nations Environment Programme (UNEP), *GEO Year Book*, 2007.
48. *Le Monde*, 8 July 2006.
49. *Nature*, 28 September 2009.
50. L'Atlas de l'Environnement, *Le Monde Diplomatique*, 2007, 50.
51. *Liberation*, 5 April 2007.
52. yahoo.fr, 26 July 2008.
53. *Nature*, Vol 439, January 2006, 16.
54. *Le Soir*, 3 April 2007.
55. *Science et Avenir*, March/April 2007, 53.
56. *Le Monde*, 18 March 2008.
57. *Ice and Snow*, 2006.
58. *Sciences et Vie*, No. 150, March/April 2007, 9.

59. *International Herald Tribune*, 2 October 2007.
60. *Le Monde*, 20 May 2009.
61. *Sciences et Avenir*, March/April 2007, 9.
62. Ibid., 35.
63. UNDP, *Human Development Report, 2007–2008*, 102.
64. Ibid., 9.
65. *Common Dreams*, 30 January 2004.
66. 'Effets des changements climatiques dans les Tropiques: le cas de l'Afrique', *Alternatives Sud*, Vol. XIII, No. 2, 2006, 85.
67. *Le Monde*, 7 April 2007.
68. *Le Monde*, 10 September 2007.
69. Anthony Nyong, *Alternatives Sud*, Vol. XIII, No. 2, 2006.
70. *Le Soir*, 18 October 2007.
71. UNDP, *Human Development Report, 2007–2008*, 9.
72. *Le Soir*, 9 April 2007.
73. *New York Times*, 11 May 2009.
74. *Foreign Policy*, December 2007–January 2008, 33.
75. *Les Clés de la Planète*, Toulouse: Milan Presse, April 2007, 44.
76. Eudald Carbonell, *El Nacimiento de una nueva Civilización* (in Catalan).
77. *Guardian*, 15 June 2006.
78. *Le Soir*, 17 September 2007.
79. *Le Monde*, 24 October 2007.
80. *New York Times*, 8 March 2008.
81. *Le Monde*, 30 May 2007.
82. *Los Angeles Times*, 18 January 2008.
83. *Le Soir*, 9 June 2008.
84. 'Géopolitique de la diversité et développement durable', *Alternatives Sud*, Vol. XXIII, No. 2, 185–96.
85. UNDP, *Human Development Report, 2007–2008*, 4.
86. *Le Monde*, 13 November 2007.
87. *International Herald Tribune*, 27 June 2008.

CHAPTER 3

1. *Sciences et Avenir*, March/April 2008, 16.
2. F. Ruddiman, *Plows, Plagues and Petroleum – How Humans took Control of Climate*, Princeton: Princeton University Press, 2005.
3. *Le Monde*, 15 July 2007.
4. *Louvain*, No. 167, March 2007.
5. *Le Monde*, 2 June 2007.
6. L'Atlas de l'Environnement, *Le Monde Diplomatique*, Paris, 2007.
7. Cited by *Courrier International*, 20–26 September 2007.
8. *International Herald Tribune*, 25 October 2007.
9. Quoted by *Le Monde*, 22 January 2006.
10. *Le Soir*, 7–8 June 2008.
11. *New York Times*, 29 June 2009.
12. *Le Soir*, 23 January 2008.
13. *Newsweek*, 16–23 April 2007, 44.
14. Ibid., 53.
15. Ibid., 62.

16. Ibid., 88.
17. Ibid., 44.
18. Ibid., 89.
19. Ibid., 53.
20. Ibid., 63.
21. Ibid., 72.
22. *Jakarta Post*, 24 June 2008.
23. *Le Monde*, 30 May 2007.
24. Cited by *Le Soir*, 10–11 March 2007.
25. *Le Soir*, 10–11 March 2007.
26. *Sciences et Vie*, January 2008, 24–6.
27. Anthony Giddens, *The Politics of Climate Change*, Cambridge: Polity Press, 2009.
28. *Le Soir*, 12 September 2007.
29. 'El Negocio de Contaminación', *ALAI*, 28 December 2007.
30. *Newsweek*, 16–23 April 2007, 46.
31. Ibid., 54.
32. Ibid., 55.
33. Ibid., 64.
34. Ibid., 46.
35. *Le Monde*, 21 September 2007.
36. Enrique Leff, 'Géopolitique de la biodiversité et développement durable', *Alternatives Sud*, Vol. XIII, No. 2, 196.
37. *Financial Times*, 30 October 2007.
38. *Wall Street Journal*, 10 October 2007.
39. *World Rainforest Bulletin*, 2 June 2007.
40. Le Monde, 3 July 2009.
41. *Newsweek*, 16–23 April 2007, 21.
42. David Adam, *Guardian*, 8 January 2008.
43. Focus on the Global South, electronic bulletin No. 135, December 2007.
44. *International Herald Tribune*, 11/12 August 2007.
45. *Financial Times*, 25 January 2008.
46. Fidel Castro, 'Se intensifica el Debate', *Granma*, 10 May 2007.
47. 'Climat et Développement, une articulation souhaitable', *Alternatives Sud*, Vol. XIII, No. 2, 2006, 147.
48. In 'La naturaleza frente a la tormenta global', *ALAI*, 10 October 2007.
49. Ibid.

CHAPTER 4

1. J.W.B. Vidal, *Brazil-Civilização suicido*, Brasilia, 2002, 25–8.
2. Grain, www.grain.org.
3. *Ecoactif*, 17 July 2007.
4. J. Berthelot, 2009, 11.
5. Arun Agrawal, 2005.
6. Anne Priéur-Vernat and Stéphane His, *Les biocarburants dans le monde*, Paris: IFP, 2007.
7. Maurice Luneau, *La documentation française*, Paris, 1982.
8. Ibid.

9. Esterification is a chemical reaction beetween an oil and an alcohol which produces ester from glycerine and fatty acids.
10. J.D. Pellet and E. Pellet, *Jatropha curcas, le meilleur des biocarburants*, Paris: Favre, 2007.
11. www.naturavox.fr/article.php3?id_article=2923.
12. For atmospheric pollution and greenhouse effects see *Que sais-je?* No. 2667 under 'Environment'.

CHAPTER 5

1. Edivon Pinto and Marleen Melot, *O mito do biocombustiveis*, Pastoral Land Commission, National Conference of Brazilian Bishops, Brasilia, 20.
2. Horacio Martins de Calvalho, 'La expansión de la oferta de etanol', www.alainet.org/active/19020.
3. *América Economía*, 2 April 2006.
4. Addison Roberto Gonçalves, *Terra Econômica*, 14 March 2007.
5. Conferência nacional popular sobre Agroenergia, Curitiba, 28–31 October 2007, São Paulo, Movimento dos Trabalhadores sem Terra, et al.
6. *La Jornada*, 13 December 2007.
7. *Argenpress*, 19 April 2007.
8. *Brempunkt*, 237, September 2002, 23.
9. D. Parizel, 'La Menace des carburants agro-industriels', *Nature et Progrès*, Jambes (Belgium), 2008, 34.
10. Alexandre Koos, *Le Monde*, 13 May 2008.

CHAPTER 6

1. Martin Lynn, *Commerce and Economic Change in West Africa – The Palm Oil Trade in the Nineteenth Century*, Cambridge: Cambridge University Press, 1997.
2. M. Kindela, *Congo Vision*, 17 April 2007.
3. www.laconscience.com/article.
4. *International Herald Tribune*, 31 August 2006.
5. Abet Nego Tarigan, Sawit Watch.
6. World Bank, *World Development Report*, 2008.
7. *Agriculture and Agri-food Canada*, 17 April 2007.
8. www.malikounda.com/nouvelle_voir.php?idNouvelle=5102.
9. J.D. Pellet and E. Pellet, *Jatropha Curcas, le meilleur des biocarburants*, Paris: Favre, 2007.
10. 'Production Practices and Post-Harvest Management in Jatropha', contribution of Lalji Singh, S.S. Bargali and S.I. Swamy from the Indira Gandhi Agricultural University, Raipur, to the Biodiesel Conference towards Energy Independence: Focus on Jatropha, 266.
11. A CER is a certificate guaranteeing that a project generates the equivalent of 1 tonne of CO_2 less than the same project emitted with traditional methods.
12. Bakima, in www.laconscience.
13. www.malikounda.com.
14. Syntheses and Conclusions of the International Conference on Issues and Prospects of Agrofuels for Africa, Ouagadougou, Burkina Faso, 27–29 November 2007.

15. A. Brew-Hammond and A. Crole-Rees, Paper presented to the Regional Conference on *Jatropha, outil de lutte cvontre la pauvreté en Afrique*, Bamako, Mali, 16–17 January 2006.
16. Mersie Ejigu, 'L'Afrique a besoin d'energie renouvelable', Point de vue, *Le Monde*, 22 June 2005.

CHAPTER 7

1. Sarojeni V. Rengam, sarojeni.rengam@panap.net and panap@panap.net, www. panap.net.
2. D. Parizel, 'La Menace des carburants agro-industriels', *Nature et Progrès*, Jambes (Belgium), 2008, 19.
3. Alejandra Parra, Rada, Network for Environment Rights Action.
4. Robert Jackson et al., in Chris Lang, http://chrislang.org.
5. George Monbiot, *Guardian*, 8 December 2005.
6. Danae S.M. Maniatis, 'Ecosystem Services of the Congo Basin Forests', Oxford University, 2007, http://globalcanopy.org/themedia/Ecosystem%20 Services%20CB.pdf.
7. Ökobilanz von Energieproduckten: Ökolische Bewertung von Biotreibstoffen. Final report, April 2007. Study commissioned by the Federal Office of the Environment and the Federal Office of Agriculture. Empa, département Technologie.
8. For a general analysis and critique of agrofuels, see the *Independent*, 17 September 2006.
9. *Le Monde*, 29 November 2008.
10. http://mwcnews.net/content/view/14507/235/.
11. www.wrm.org.uy.

CHAPTER 8

1. *Nourrir les Hommes*, Ed. du Cep and Paris, Office général du Livre, Brussels, 1963.
2. Jean Ziegler, Special Rapporteur of the United Nations on the Right to Food, *L'Empire du Chaos*, Paris: Fayard, 2005.
3. 'A Precarious Existence: The Fate of Billions', in Samir Amin, *World Poverty, Pauperization and Capital Accumulation. Monthly Review*, October 2003.
4. Samir Amin, *Le capitalisme et la nouvelle question agraire*, Dakar: FTM, 2004.
5. *National Geographic Magazine*, February 2007, 25.
6. *Rebellion*, November 2008.
7. Laetitia Clavreul, *Le Monde*, 16/17 March 2007.
8. H.C. Binswanger, 'Impératifs économiques et écologiques d'une politique agricole à long terme', *Horizons et Débats*, 7th Year, No. 11, 26 March, 2.
9. Eco Portal.net, 21 September 2007, http://ecoportal.net/.
10. América Latina en Movimento, *ALAI*, 17 July 2007.
11. Center for Public Integrity, 24 September 2007.
12. World Rainforest Movement ,*WRF*, 2 June 2006.
13. *WRF*, 6 June 2006.
14. *Palme, finance et pouvoir politique*, June 2007.
15. *Le Monde*, 20 October 2007.
16. Solidarité: http://solidarite.asso.fr, 30 May 2008.

CHAPTER 9

1. *New York Times*, 6 February 2007.
2. *Science et Vie*, No. 1086, March 2008, 56.
3. Michel Destot et al., *Energie et Climat*, Paris: Jean Jaurès Foundation and Plon, 2006, 45.
4. *Le Monde*, 2 June 2007.
5. *Le Monde*, 25 November 2008.
6. *L'Echo*, 25 May 2009.
7. Guy Cloes, 'Guide des Energies renouvelables', Wallon Regional Ministry, Brussels, 2007.
8. *Le Soir*, 4 June 2008.
9. *New York Times*, 29 November 2008.
10. *Le Soir*, 24 September 2006.
11. www.futura-sciences.com.
12. Since 1992, in the Gironde in France, geothermy has been reheating a sturgeon fish farm and it is expected that caviar will soon be produced.
13. Enrique Dussel, *L'Ethique de la Liberation*, Paris: L'Harmattan, 2004.
14. François Houtart and Geneviève Lemercinier, *Culture et Energie*, Paris: L'Harmattan, 1982.

Index

Compiled by Sue Carlton